The Atlas of Bird Migration

Tracing the great journeys of the world's birds

General Editor Jonathan Elphick

Foreword by Thomas E. Lovejoy

A Firefly Book

Published by Firefly Books Ltd. 2011

Copyright © 2011 Marshall Editions

First printing

Publisher Cataloging-in-Publication Data (U.S.)
Elphick, Jonathan.
   Atlas of bird migration : tracing the great journeys of the
world's birds / Jonathan Elphick, general editor ; foreword by
Thomas E. Lovejoy.
Originally published 2007.
[   ] p. : col. ill. ; cm.
Summary: Explanation of flight techniques, navigation, feeding
and biology of migrating birds as well as environmental
threats to migrating species and conservation initiatives. Maps
trace migration routes of over 100 species of birds divided by
geographic regions and a supplementary catalogue details the
routes of 500 additional species.
ISBN-13: 978-1-55407-971-1 (pbk.)
1. Birds--Migration. I. Lovejoy, Thomas E. II. Title.
598.156/8 dc22   QL698.9.E47 2011

Library and Archives Canada Cataloguing in Publication
   Atlas of bird migration : tracing the great journeys of the
world's birds / general editor, Jonathan Elphick.
Includes bibliographical references and index.
ISBN 978-1-55407-971-1
   1. Birds--Migration. 2. Birds--Behavior.
3. Birds--Migration--Maps. I. Elphick, Jonathan
QL698.9.A89 2011   598.156'8   C2011-901166-2

Published in the United States by
Firefly Books (U.S.) Inc.
P.O. Box 1338, Ellicott Station
Buffalo, New York 14205

Published in Canada by
Firefly Books Ltd.
66 Leek Crescent
Richmond Hill, Ontario L4B 1H1

Printed in China

Right: bird eagle in flight

Contents

Foreword

by Dr. Thomas E. Lovejoy, the H. John Heinz III Center for Science, Economics and the Environment

The spiraling "kettle" of large birds was dramatic in the Egyptian sky as hundreds of European White Storks, their white plumage gleaming in the desert sun, rode thermal currents to gain altitude on their southward migration. This vivid observation in the fall of 1962 made real to me the challenge that large heavy migrating birds face. They simply aren't built to make the trip on a direct course flapping basis, but must purloin energy from wind currents. This leads to what in a sense are inventive migratory pathways, ones that avoid large water bodies like the Mediterranean and seek out the kinds of landforms that generate thermal currents.

Such tortuous pathways and the myriad physical feats that all long-distance migrants face twice a year are made worthwhile by the advantages of summertime habitat with its great flush of food sources, plus the similar advantages of wintering grounds. While humans compete in the New York marathon, what goes on in the skies overhead is extraordinary in comparison. Small wonder, then, that anyone who has ever given the matter a thought is inevitably intrigued by these colossal avian achievements.

Now, as I pause to think about migratory birds, it comes as less of a surprise than I would have believed at the outset to realize the extent to which they have intersected my existence. Nothing can ever erase the thrill I experienced the first time I learned as a schoolchild to use binoculars and witnessed a spring migration: the tidy crisp pattern of the Black-throated Blue Warbler, and the elusive Worm-eating Warbler, which accommodated by hanging upside-down in its treetop tower so we could see the telltale fieldmarks of stripes on its crown. Today, close to 40 years later, the call of a Wood Thrush, or the brilliant orange plumage of a Prothonotary Warbler, continues to elicit an equal thrill.

Behind these impressive feats and the delight they provide for millions of birdwatchers lies fascinating biology, much of it still to be revealed and understood. Biologists largely comprehend the offset breeding season of the few thousand Eleonora's Falcons, which permits them to take smaller migratory species as food for their young prior to the falcons' own migration from Mediterranean islands to Madagascar. But how Ruby-throated Hummingbirds, with hearts that beat 500 times a minute and energy demands of gargantuan proportions, are able to migrate across the Gulf of Mexico is still a mystery. The wintering grounds, ecology, and behavior of many migratory species are still largely unknown. We have little ken of the biological mechanism that causes a bird to migrate, or not. In Europe starlings are migratory, but the overwhelming majority of those introduced in the United States are not.

Like other elements of the natural world, birds have not always had benign interactions with people. In the United States it used to be considered sport to hunt migratory birds, particularly birds of prey. That, happily, is history, thanks to the efforts of Rosalie Edge of the Hawk Mountain Sanctuary (who once managed to make time for this then-fledgling conservationist) and similar individuals who built the National Audubon Society. Regrettably, migratory birds are still hunted in Europe.

Sometimes the problems of migratory birds are relatively easily solved—by the protection of a staging area of migratory shorebirds, for example. Driven as much by the cost of downed aircraft as by the deaths of the birds with which they collide, Israel now has a policy whereby a glider pilot travels with migrating flocks as they pass through in order to keep military and other aircraft aware of potential hazards. A flight in that glider brought me as close as possible to understanding what the life of a thermal glider must be like.

More difficult to deal with are the problems of many declining North American migrants, such as the Kentucky Warbler, so well presented in John Terborgh's Where Have All the Birds Gone? Part of the problem is lack of scientific information. That is being tackled by various organizations, including the Smithsonian Institution and the National Fish and Wildlife Foundation. More complex are the causes which, at least in broad brush, are clearly responsible, including changes in use of the land both in the summer and in the winter range.

Migratory birds do, in the end, integrate and reflect environmental change on the broadest of scales. The sobering truth is that whether future generations can enjoy this truly remarkable phenomenon of nature will be more a measure of whether humanity will have achieved sustainable development than anything else.

Thomas E. Lovejoy

Using this book

Bird migration is a complex subject, and knowledge of patterns of movement for some species and in many areas of the world is, at best, sketchy. The information given in this book is accurate, as far as is currently known, but the subject remains one about which there is much yet to be discovered.

Within each chapter, species are arranged according to the systematic order, which is approximately the order in which they evolved. This scientifically accepted system has the advantage over other methods of organization that birds that are related are grouped together.

Species have been chosen for inclusion on one of two major criteria: their migrations are typical of one or more patterns of movement and/or there are interesting stories associated with their migrations.

Common names for birds tend to vary and are often a matter of choice or editor's preference; those used in the book reflect general usage as far as possible.

THE MAPS

One color is used for each individual bird's breeding and wintering areas, as well as for the relevant migration arrows, dotted dispersal limit lines, and any staging posts that it may use on migration.

❶ Arrows on the map show the broad direction of travel. A broken arrow indicates that the route is suspected but not proven. If a route is not known, the broken arrow is marked "Route not known." Unless indicated otherwise, birds travel out from and back to their breeding grounds by the same route. Where movements are more dispersive than truly migratory, such as with some sea birds that breed in clearly defined areas then drift the oceans until the next breeding season, the maximum limits of the birds' dispersal are indicated by colored areas.

❷ The colored areas of the maps show the breeding and wintering ranges of one or more birds from a group. If the ranges of two birds overlap, different colored stripes have been used. An equal number of stripes in an area indicates that the two birds are equally widespread there.

❸ Where a bird's breeding and wintering ranges overlap, black dotted lines indicate the northern and southern limits of the ranges.

❹ THE CALENDARS

For each land bird included on a map, a calendar indicates the months of migration (outward and return) and the months that birds spend on the breeding grounds. Calendars for some sea birds indicate months on the breeding grounds and months away from them (non-breeding).

The color used for a bird on the map is repeated in the central part of the calendar and the arrows in the concentric breeding and migration (or non-breeding) circles.

Months are indicated around the outside of the calendar. The points on the edge of the calendar wheel mark the ends of the months.

SYMBOLS AND ABBREVIATIONS

- ✪ **Hot spot** Place where migrants can be seen in large numbers
- ◯ **Staging post** or **stopover area** Places to stop to feed en route
- ✖ **Threats** to migrants
- ♂ **Male**
- ♀ **Female**

- **N** North/northern
- **S** South/southern
- **E** East/eastern
- **W** West/western

- **Is.** Island(s)
- **L.** Lake
- **U.S.** United States

- **max.** maximum
- **pop./pops.** population(s)
- **sp./spp.** species

- **NNP** National Nature Park
- **NP** National Park
- **NR** Nature Reserve
- **NWR** National Wildlife Refuge
- **WR** Wildlife Refuge

Throughout the book, the word billion refers to one thousand million

THE FACT FILES

Basic facts are given for birds that feature prominently on a spread. Alongside the bird's common and Latin names, weight, wingspan, and the length of its migratory journey is a silhouette of the bird inside a grid, to give a broad indication of its length from beak to tail. Five grids have been used, but in each case one square of every grid represents 4 inches (10 cm), so that larger birds sit on grids containing more squares than smaller ones.

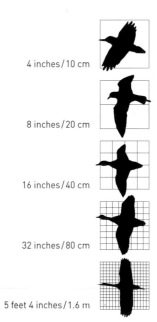

4 inches/10 cm

8 inches/20 cm

16 inches/40 cm

32 inches/80 cm

5 feet 4 inches/1.6 m

Essential facts about many other birds mentioned in the text, along with those of many more migratory species, including their breeding and wintering ranges and length of journey, can be found in the Catalog of migrants on pp. 166–172.

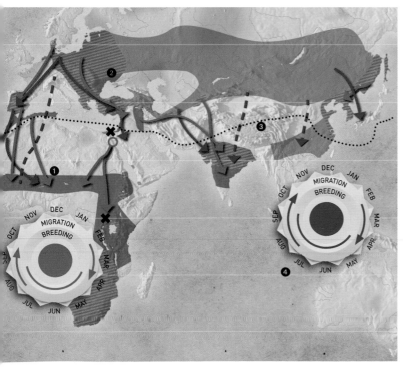

Birds on the move

Migration is probably the most awe-inspiring natural phenomenon. What it lacks compared with the enormous power of the weather, an earthquake, or a volcano, it makes up for in romance—a small bird pits its wits against the elements and accomplishes, as routine, a journey that is truly superhuman. Outlandish theories have been proposed to explain the seasonal ebb and flow of bird populations, including that the birds went to the moon, were transformed into other species, or spent the winter in the mud of lakes or ponds. The knowledge of what really happens is, in many ways, no less fantastic.

Over the centuries, people have struggled to understand birds' migrations and how they manage them year after year. How can they fly such tremendous distances without becoming lost or so exhausted that they die? How have today's complex migration patterns evolved from what were once, presumably, simple movements? And what particular problems are migratory birds facing now and are they coping? Although many parts of the mystery have been unraveled, there is still much that remains to be discovered about this fascinating phenomenon.

Barnacle Goose in flight at sunset around Mersehead in Scotland, where they migrate for the winter from their breeding colonies on Svalbard.

How migration evolved

Shifting continents and the ebb and flow of ice sheets affect migration patterns and pathways.

The initial evolution of migration is easy to imagine, driven by the changing seasons. Northern summers are warm and winters cold, so a bird that thrives in warmer areas will benefit by moving south in winter. If it stays in the northern hemisphere, the weather will be less severe; if it crosses the equator, it can enjoy the southern summer. There is likely to be an area to the south where the species could live year-round, but northern breeders may be forced onward by competition from residents of the same species. This results in a long migration to an area where the summers may be too hot to breed, but where winter conditions are ideal.

Many other, less immediately obvious, factors have, over geological time, caused birds to alter their movements and influenced the evolution of new species. In the long term, continental drift affects distribution and migration patterns. Some 50 million years ago, there were many bird species that would be identifiable today, at least to group level. Then, the pattern of the continents was different. South America was some 600 miles (1,000 km) from North America, and India was 1,200 miles (2,000 km) away from the rest of Asia, but Africa was close to Eurasia on a front 3,000 miles (5,000 km) long. It is no coincidence that the most complex migration systems are those that have developed between Eurasia and Africa, while those between North and South America, and between Asia and India, are simpler.

The long-term factor that has influenced migration most, however, has been the succession of ice ages that have affected the world. There have been about half a dozen very cold periods over the last two million years—two of which occurred in the last 150,000 years. In times of remission, temperatures fluctuated widely in cycles lasting between 50,000 and 100,000 years. Research suggests that some of these changes have been rapid— over tens of years, rather than hundreds—and this must have put a premium on flexibility in migratory strategy.

The cycle of ice ages has pushed the habitat that suits a particular bird species backward and forward. This may affect both its summer and winter quarters, but for a long-distance migrant, the effects are more noticeable in the northern, summer, breeding area. Birds can cope with gradual change because individuals, from year to year, have to modify their journeys only on the edges of the range. Over 1,000 years or so, ranges of birds may shift over the globe. But since the change is slow, the birds keep to their general movements over thousands, even millions, of years.

❶ MORE THAN 100,000 YEARS AGO, the world was warmer and the polar ice cap smaller than today. This had a significant effect on ancestral bird species in both the Old and New worlds: they extended into the far north. In North America, an ancestral warbler species also extended across the continent from Alaska in the west to Newfoundland; in Europe, the west–east range was from the Atlantic coast to the Urals.

Audubon's Warbler

Myrtle Warbler

THE NORTH AMERICAN MYRTLE AND AUDUBON'S WARBLERS, once regarded as separate species, are considered forms of the same species, the Yellow-rumped Warbler. There are different races of both forms from various parts of their range, and these can be directly related to the way in which glaciations have affected North America.

The Eurasian Common and Lesser whitethroats are two distinct, though obviously related, species. Speciation occurred two glaciations ago, and at the last glaciation, Lessers were found only in the east, while Commons remained in both the east and the west.

Common Whitethroat

Lesser Whitethroat

❷ AT ITS FULLEST EXTENT, an "intermediate" glaciation laid ice sheets so far south that the North American ancestral species was not only pushed out of the whole of its earlier range, but was also fragmented. In Europe the species was forced into two areas—the Balkans and Near East, and Iberia. This situation remained stable for thousands of generations.

❸ DURING AN INTERGLACIAL, the extent of the ice was similar to that of today. This allowed the birds to regain lost ground. As this happened, sibling species began to move closer to one another, and in Europe overlapped at the north of the range. Because they had been separated for so long, the populations had differentiated enough not to interbreed readily. The Myrtle and Audubon's warblers at this stage were still similar, but the Common and Lesser whitethroats were separate species.

❹ AT THE HEIGHT OF THE LAST GLACIATION, 15,000 years ago, ice pushed the birds south again. In Europe, Common Whitethroats retreated into both of their former refuges, but Lessers moved to the Balkans only.

In America, the situation in the east was simple—only Myrtle Warblers were there. In the west, some Audubon's Warblers retreated into Mexico; northern populations were joined by Myrtle Warblers.

- ■ Ancestral American warbler
- ■ Audubon's Warbler
- ■ Myrtle Warbler
- ■ Ancestral Eurasian warbler
- ■ Common Whitethroat
- ■ Lesser Whitethroat

❺ TODAY THE POPULATIONS have spread back north. Common Whitethroats have an eastern and a western race coinciding with the populations in the two former havens. In North America, there are two distinct forms of Myrtle Warbler and four ot Audubon s trom Alaska to Mexico.

Patterns of migration ①

Half the world's bird species migrate, yet each journey is unique.

From movements of a few hundred yards to flights that circumnavigate the globe, from north to south and east to west, birds' migratory journeys are as varied as the species that undertake them. Defining types or patterns of migration is not easy. Nonetheless, some trends can be discerned in these unique journeys, although none of these patterns is exclusive, and one species may fall into more than one of the categories.

North to south and staying behind

Probably the easiest migratory pattern to understand is long-distance north–south migration. Barn Swallows, for example, can live in most of North America and Eurasia in summer, feeding on the seasonal crop of flying insects. Over most of their breeding range, the winter weather is so severe that there is little chance of the birds surviving, so they simply move south to areas where they can continue to find food.

This is not as clear-cut as it sounds, however. Birds have no reason to move if the climate is equable year-round and has no effect on the availability of food. Where a species is widespread, breeding over a broad band of latitudes, individuals from the northern parts of the range may have to fly far to the south. Those from the southern parts of the range—where winter conditions are less severe—by contrast, may not need to move at all. Barn Swallows in southern Spain are resident in their breeding range all year round, as are the Killdeer of the Gulf States of North America.

In many species in which this pattern occurs, another migratory phenomenon may be overlaid: that of partial migration, in which some birds of the population in an area migrate and others do not.

❷ ARCTIC TERNS are champion long-distance migrants. They nest during the Arctic summer and start to fly south as early as August. They follow the coastline of Africa to reach the Antarctic in November and drift with the winds that blow east. After the three months of the southern summer, the adults begin to retrace their path north.

❸ REDWINGS are east–west migrants, breeding across Russia and northern Europe. As the northern summer draws to a close they start to move west and south, flying by night to reach the wintering grounds of western and southern Europe in October. Birds may winter in different areas each year: those banded in Britain one winter have been found in Greece, Turkey, and the country of Georgia the next.

❶ IN SOUTH AMERICA, Vermilion flycatchers are south–north migrants. They breed in the southern summer, between September and February, on the pampas of Uruguay and northern Argentina. In the fall they fly north to the savannas of Brazil and Colombia, where they join a resident population from March to August.

4 RED KNOTS are long-distance north–south migrants, breeding in the short Arctic summer. A young bird's first flight may be over the forests of Asia to traditional shorebird staging posts in Southeast Asia. Many red knots fly over the Pacific Ocean and South China Sea to the Australian coast, then cross the continent to winter on its southern and western coasts.

5 BLUE GROUSE of the Rocky Mountains are altitudinal migrants, moving vertically about 1,000 feet (300 m). They breed at low altitudes in deciduous woodland clearings, feeding on berries and insects. In winter they move up the mountains to coniferous woodland. Here they feed on pine needles which, though abundant, are not nutritious enough for the birds to be able to breed and rear their young on them.

6 SNOW GEESE breed in Arctic Canada and Alaska, and as far west as northeastern Siberia and as far east as western Greenland. Their migration is punctuated by periods spent at traditional wetland staging posts.

In winter they feed intensively on coastal marshes and farmland: reserves are vital to sustain them for the first week or two back in the Arctic.

Patterns of migration ②

Where partial migration occurs, the juveniles often migrate while the adults stay put. This strategy may remain stable for generations, with short-term advantages for one of the two groups outweighed overall in the long term. If there are distinct advantages for one group for several consecutive years, the other dies out and the pattern of the survivors becomes the "standard."

Coasts and mountains

Many species have developed migrations that are more east–west than north–south. Usually these are birds that are taking advantage of the better winter climate provided by the sea at the edge of a continent. In Europe chaffinches move east from Scandinavia to Britain (they are also partial migrants with the males staying behind). In North America most White-winged Scoters breeding in western Canada and Alaska winter on the eastern seaboard. Shorebird migration is also dominated by the need to stay close to water, with routes hugging coastlines and major stopovers on estuaries.

Chaffinches in some parts of their range also typify another strategy: altitudinal migration. Birds breeding in the Alps may move horizontally, but they also move vertically to escape the winter weather on the upper slopes.

For many species, potential wintering habitats are vast. Those with special needs, by contrast, have often evolved specific migrations, with routes and places to stop and feed en route passed on from parent to young. Families of geese and swans travel and winter together, and the entire population of Whooping Cranes travels from Canada to a restricted winter site in Aransas, Texas.

Southern hemisphere migrants

The predominant migratory pattern in the southern hemisphere might be supposed to be south–north. To a certain extent this is true: many birds breeding in the temperate latitudes of Australasia, South America, and Africa migrate to the tropics and subtropics in winter. Since there is so much less land in the southern hemisphere than in the northern, and so much more ocean, many of the birds that breed in the south are sea birds. While their migratory patterns are also dominated by the need to find food, winds and ocean currents play a more vital role than temperature in determining the direction and extent of migrations.

FOX SPARROWS are leapfrog migrants. This migratory pattern occurs when a species that is resident in an area develops migratory populations that have to overfly the area inhabited by the residents. They are unable to share the area since the competition for resources would almost certainly favor the residents with their additional local knowledge.

There are advantages for both populations: the residents are able to breed early; the migrants can stay on the wintering grounds until conditions are right for them to nest in their own breeding area.

■ race *unalaschcensis*
▫ race *townsendi*

This pattern develops as a species extends its range poleward. The different Fox Sparrows have spread along the Pacific coast from wintering areas as far south as Mexico to breeding populations on the Aleutian Islands. These populations are far enough apart to have developed into recognizable races. The race *townsendi* breeds on the coast of Canada and winters over the border in the northern U.S.; the Alaskan breeders of the race *unalaschcensis* leapfrog these birds to winter in California, Nevada, and Idaho. (Other populations are not shown.)

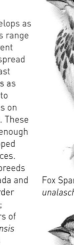

Fox Sparrow race
unalaschcensis

race *townsendi*

Evening Grosbeak

♀

♂

EVENING GROSBEAKS
are becoming partial
migrants. On the northern
fringes of their range the
extent of movement away
from the breeding areas
is determined by food
supplies. In years when
food was plentiful they
were able to stay; in other
years they may have had
to travel south. Winter
feeding at bird feeders—
they are partial to black
sunflower seeds—has
altered this situation
in many areas. Today,
male Evening Grosbeaks
often stay, even in the
northernmost areas, while
females move southward.

The range of these
birds has expanded rapidly
eastward since 1920, when
they first bred in Ontario.
Now they breed on the
eastern seaboard in Nova
Scotia, 1,250 miles (2,000
km) away. There are also
records of these birds
having reached Europe.

MANX SHEARWATERS
(above), in common
with other sea birds,
often undertake lengthy
migrations because ocean
resources are seasonal.
These birds spend the
summer on the remote
islands off the coasts of
western Europe, where

they breed in burrows.
In the fall they leave the
colonies to winter off the
eastern coasts of South
America, from Brazil
to Uruguay. Several
recoveries of banded birds
show that even young
birds can complete this
6,000-mile (10,000-km)
journey in less than three
weeks. The return journey
is more leisurely, with
birds heading north to the
Caribbean, then using
the currents of the Gulf
Stream to help them
across the Atlantic.

When to travel

Flying by day and by night both bring their own rewards.

Although many birds show a preference for day or night flight, for most species migration regularly involves travel at both times. The major differences between day and night travel are the methods the birds use to navigate (see pp. 30–33) and the temperatures they experience en route.

Migrating by day

There are some species whose main flights are wholly confined to daytime because they rely on the rising air from thermals (see pp. 16–17). Migrants that depend on this means of travel rarely stir until some three hours after dawn and often stop the day's flight in the early to mid-afternoon. Soaring flight is so energy-efficient that these birds do not need to feed much, if at all, and can spend the rest of the day and night roosting.

Many species of relatively short-distance migrants are perceived to migrate only in the early morning, although whether this is the case is still open to debate. Some birds appear regularly on radar screens at dawn. Radar devices have also, however, picked up echoes of birds descending and continuing to fly at low level. Whether these are the same birds is not certain. If some birds do indeed confine their migrations to the hours immediately after dawn, then the most likely candidates are such birds as Meadow Pipits, Sky Larks, and finches. A four- or five-hour flight would allow them to cover a distance of up to 95–125 miles (150–200 km). At this rate, many birds could achieve their full migration in four or five stages. They would also not need well-developed nocturnal navigation skills.

Barn Swallows and Sand Martins roost in reedbeds at night and migrate only during the day. Their close relatives the House Martins, on the other hand, do not roost in vegetation and may continue to fly through the night unless they are able to find and occupy a spare nest in a colony.

Nocturnal migrations

Birds rarely fly only at night, although nighttime migrants are common. Land birds are unable to come down on the sea, and since many nocturnal migrants make ocean crossings, they are forced to continue to fly by day. Shorebirds crossing the Pacific Ocean to winter on its islands, or those continuing on to Australasia, may fly long distances by day and night to avoid touching down.

Scandinavian and Russian breeding birds often move westward to winter in France, Spain, or Britain. Those that cross from Holland or Belgium to southern Britain have had a 160-mile (250-km) North Sea crossing; those arriving in Scotland from Denmark and Norway have flown 500 miles (800 km). Birds that are usually considered nocturnal migrants, such as thrushes, Goldcrests, and starlings, may continue to arrive until the early afternoon in the south or throughout the day into the evening in the north. Often they settle on the first land that they reach, but in less than an hour or two, have dispersed inland to find a suitable place to feed.

Flying nonstop

There are various categories of nonstop migrants. Some birds simply choose to migrate in only one or two long stages, despite the existence of suitable habitat below them. These are often small species that have put on high fat reserves for the journey at special feeding areas. The Sedge Warbler, for example, averages an additional fat load equal to 100 percent of its normal weight, with which it can probably cover more than 1,900 miles (3,000 km) in one nonstop flight of three or, perhaps, four days.

The tiny Blackpoll Warblers of the United States are also remarkable nonstop migrants. In the fall, these birds put on fat along the Massachusetts coast; some then fly out to sea to the southeast. About 36 hours later, they reach the influence of the northwest trade winds, which give them a free ride back toward the West Indies and perhaps even as far as the north coast of South America in only four days (see Wood warblers, pp. 74–75).

Many migrants were once thought to cross the Sahara nonstop. Recent research, however, suggests that some, even many, of the birds which have to cross the desert and were thought to do so without stopping may in fact land and rest during the day. It may be that as the desert heats during the day, the air above it becomes too hot for the birds' metabolism to rid the body of excess water (see Wagtails and pipits, pp. 106–7).

Finally, there are those birds that are pure flying machines, living in the air and able to fly by day or night, or both, at will. Swifts are among the champions in this respect—those that leave northern Europe in the fall are likely to remain on the wing until they return to their nest sites in northern Europe nine months later. And, at sea, terns rarely stop flying since, if they settle on the water for any length of time, they become waterlogged.

BIRDS FLYING across the face of the moon are living proof that some species, including Sanderlings (above), are nocturnal migrants.

Many moon-watchers are attracted by the romance of looking at birds seeming to float across a silver globe, but the practice also yields interesting results. The records of passes across the moon may exceed 200 per hour. For species of small birds, which are probably traveling on a front 60 miles (100 km) wide, this might mean that three million migrants an hour are passing by. The other proof of nocturnal migration comes from the birds' calls. With thrushes and other small and medium-sized birds, these often allow flocks to keep together.

Flight techniques ①

Moving on the wing means making the maximum use of energy.

Although other creatures can fly using wings of skin or skinlike membrane, bird flight depends on one unique adaptation: feathers. It is easy to overlook their special properties. As well as being excellent insulators, feathers are light, flexible, strong, and resilient. They are also highly maneuverable: birds are able to use them to control airflow in far more complex ways than aircraft use aerofoils.

It is a common misconception that a bird's wing supports the bird's weight in flight by pressing down on the air and that the upbeat simply returns the wing to a position in which it can support the bird again. In fact, the wingbeat is not nearly so simple. High pressure under the wing and low pressure above it work together to keep the bird aloft. The wing first moves down and backward, pushing the air in the same directions. It then moves up and forward, speeding the movement of air over the top of the wing, thereby reducing pressure and sucking the bird upward.

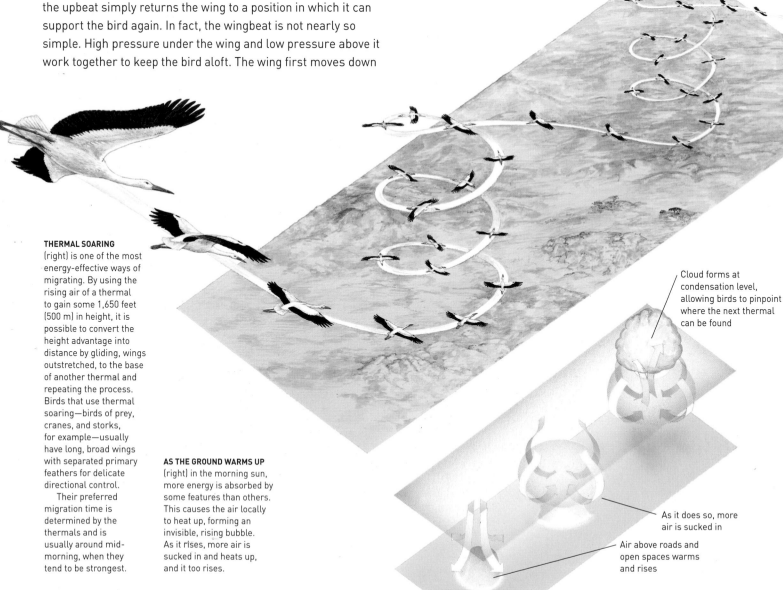

THERMAL SOARING

THERMAL SOARING
(right) is one of the most energy-effective ways of migrating. By using the rising air of a thermal to gain some 1,650 feet (500 m) in height, it is possible to convert the height advantage into distance by gliding, wings outstretched, to the base of another thermal and repeating the process. Birds that use thermal soaring—birds of prey, cranes, and storks, for example—usually have long, broad wings with separated primary feathers for delicate directional control.
    Their preferred migration time is determined by the thermals and is usually around mid-morning, when they tend to be strongest.

AS THE GROUND WARMS UP
(right) in the morning sun, more energy is absorbed by some features than others. This causes the air locally to heat up, forming an invisible, rising bubble. As it rises, more air is sucked in and heats up, and it too rises.

Cloud forms at condensation level, allowing birds to pinpoint where the next thermal can be found

As it does so, more air is sucked in

Air above roads and open spaces warms and rises

BIRDS NOT SPECIALLY adapted to soaring move forward in one of three ways: continuous flapping (right, top); bouts of flapping interspersed with periods of gliding with wings outstretched (right, center); or flapping and ballistic flight, with wings closed, known as bounding flight (right, bottom).

Continuous flapping is primarily practiced by birds with a relatively high weight compared with their wing area. They are often birds that use their wings to assist them in swimming under water. Water is denser than air so only birds with relatively small, strong wings can use them for swimming.

The best strategy for most birds with a body weight of 5 ounces (140 g) or more is to flap and glide, alternating periods of continuous flapping flight, in which the bird gains both height and speed, with periods of gliding. Gliding uses perhaps less than 1/20 of the energy needed for flapping. The length of the glide depends on the direction and speed of the wind.

Small to medium-size birds, such as warblers, finches, and thrushes, use bounding flight. The drag created as the air passes over an outstretched small, broad wing cancels out any lift the bird would obtain from gliding. These smaller birds, therefore, fold their wings and drop between periods of flapping. Surprisingly, the bird's bodies, with wings closed, generate significant lift and reduce drag.

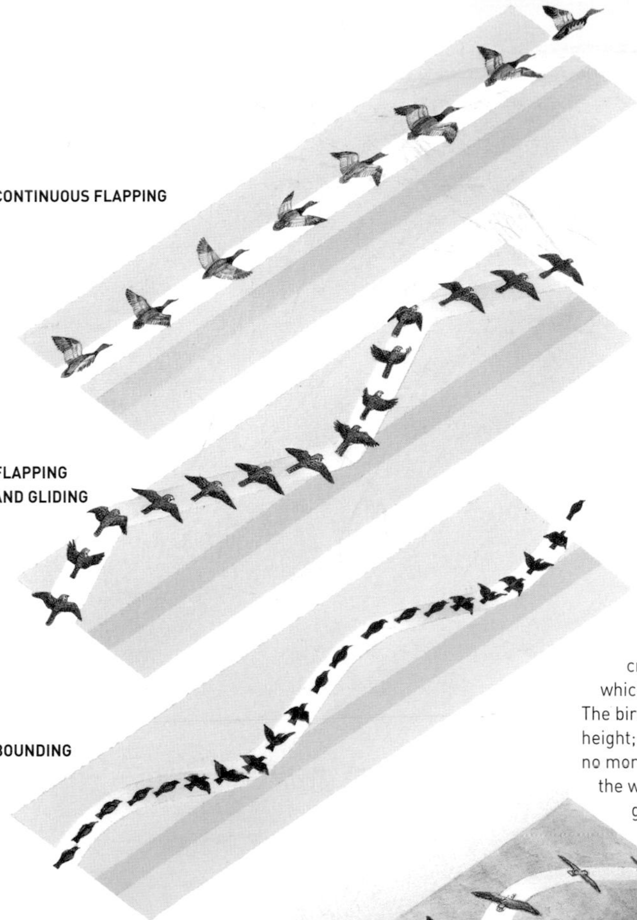

CONTINUOUS FLAPPING

FLAPPING AND GLIDING

BOUNDING

SPECIES THAT ARE ADEPT at dynamic soaring (left) make use of the friction created by the wind over water, which causes air to slow down. The bird climbs into the wind to gain height; then, when it can generate no more lift, it turns and glides with the wind, losing height but gaining ground. To enable it to make use of this technique, a dynamic soarer has long, thin wings

Wind strength usually increases with altitude

which it keeps rigidly outstretched, seemingly for hours on end. Its downwind glide in the rising air over the waves may move the bird forward 330 feet (100 m) or more before it starts to lose height.

The disadvantage of dynamic soaring is that larger birds, such as albatrosses, are becalmed when the wind is not blowing—not a common occurrence in the southern oceans these birds frequent, but a problem for any bird that drifts into the doldrums by mistake.

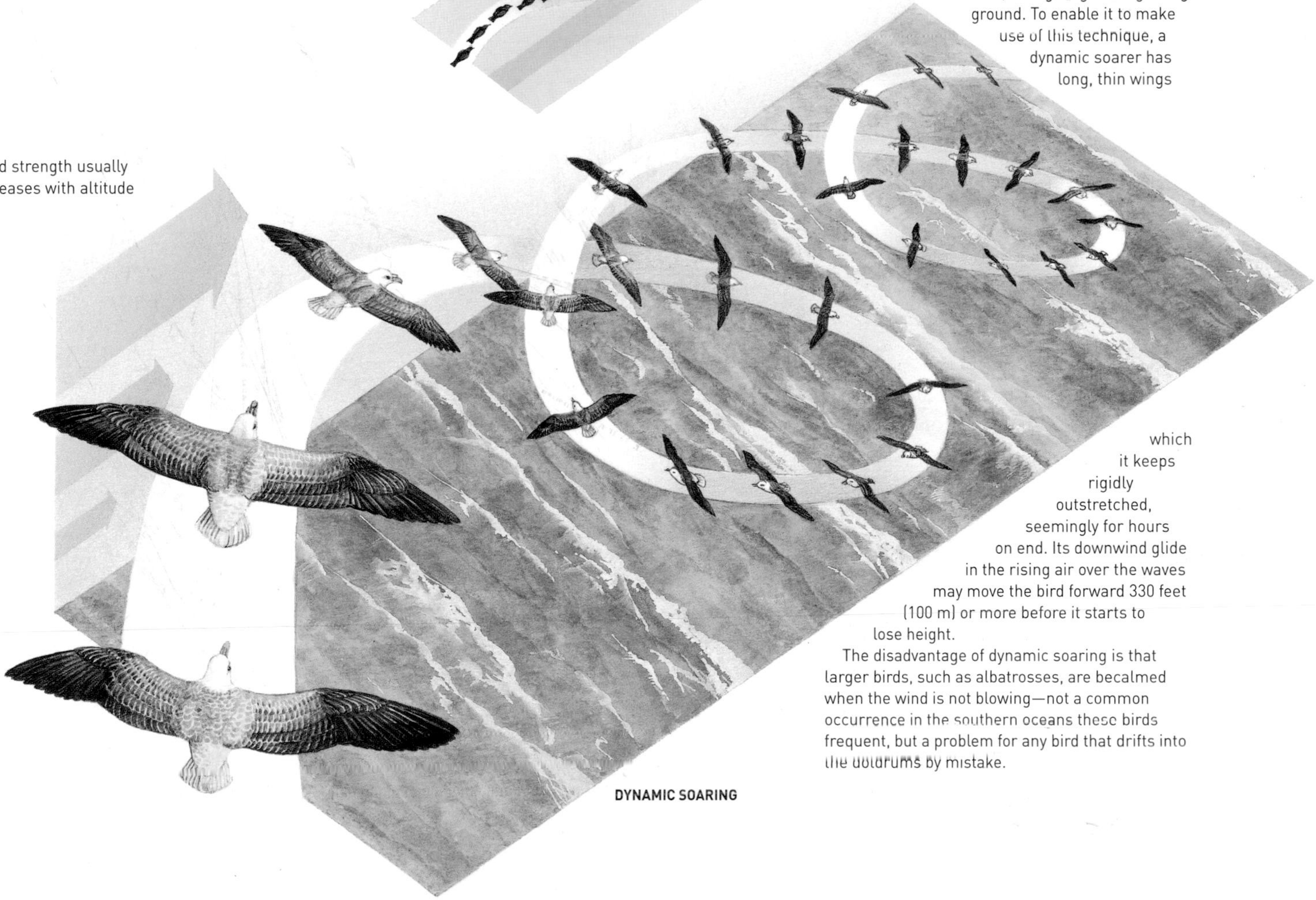

DYNAMIC SOARING

Flight techniques ②

The smallest wings are capable of the most complicated maneuvers, allowing a bird to fly forward, hover, and fly backward. The wings of hummingbirds beat at least 25 (and sometimes up to 50) times a second. Problems of scale mean that a large bird, like a swan, has no chance of beating its huge wings at anything approaching that rate or being able to maneuver so precisely. A hummingbird may weigh less than $\frac{1}{16}$ ounce (2 g), a swan a little less than 45 pounds (20 kg)—about the same as 10,000 hummingbirds.

Size, weight, and power

Swans and the smaller condors, storks, bustards, cranes, and albatrosses, at around 20–30 pounds (10–15 kg), are among the largest birds that can use flapping flight. The weight limit is determined by the efficiency with which the flight muscles provide power, which in turn is determined by the size of the muscle.

Bird weight, muscle size, and power are intrinsically linked. As a muscle doubles in width it quadruples in area, which is what gives the bird lift, but its weight increases eightfold. A bird that is twice as long as another may be carrying eight times its weight in muscle but can only generate four times as much power. Birds could not be bigger and still use powered flight unless they evolved a new type of muscle. Gliding is different, since the birds' flight muscles have only to hold the wings outstretched, so their upper weight limit is considerably higher.

Flight techniques are varied and depend to a large extent on the weight of the bird and its wing size and shape. These in turn are linked to the way in which the bird's lifestyle has evolved: it is obviously counterproductive to have long, thin wings for efficient soaring flight and live in the tree canopy chasing insects for food. Most migrant birds spend only a small proportion of their lives making migratory flights: specific adaptations for those flights, therefore, are not crucial in the species' evolution. The longer and more taxing the migration, however, the more likely it is that the bird will possess many of the features for efficient flight.

Birds can also exploit the air to the full. Takeoff, for example, is energy-inefficient, so the bird faces into the wind to become airborne and increase its speed as quickly as possible after takeoff, since flight is more efficient at a reasonable speed. Soaring birds clearly make use of the lift provided by thermals,

WALKING MIGRANTS

The need to migrate is not confined to birds that can fly. On the African plains, the pattern of rainy and dry seasons causes some Ostriches to undertake regular movements, just as, in the cold wastes of the Antarctic, Emperor Penguins trek over the ice to their remote breeding grounds.

Emus are among the most mobile of flightless birds: banded individuals have been recorded 300 miles (500 km) from the site of banding. Once the rainy season begins, they usually move to more arid areas for breeding, then return to places with a reliable water supply when the breeding season is over. Often the only land with a regular water supply has been cleared for agricultural use, with long, Emu-proof fences erected to frustrate the birds.

Emu

but other birds also use the free ride they provide to reduce their energy consumption.

The strategies that birds use for movement are, to a large extent, determined by the bird's size and shape. Most birds are able to fly using continuous wingbeats for at least a short period, to escape predators or in another emergency. The ability to glide or soar, on the other hand, is limited to birds with the appropriate adaptations. Some of these are connected with scale: flapping and gliding is clearly not an option for birds as small as starlings, sparrows, and swallows. The aerodynamic flow over the wings of small birds cannot be smooth enough for them to function properly: the drag produced at the wingtips always outweighs the lift that can be generated by the main part of the wing.

THE CHARACTERISTIC FORMATION for many flapping and gliding large birds is a V. This is an efficient way for birds to keep together. It allows each bird to see the one in front, without being impeded by the air it disturbs—particularly when the bird in front flaps and huge swirls of air are shed off the tips of its wings.

The leading bird in a V is usually an adult, unless the flock is on a routine flight, to a nighttime roost for example, when the young may take a turn.

SWIMMING MIGRANTS

First-year Razorbill

Penguins and auks regularly migrate long distances by swimming. Many auks fledge before they can fly and, accompanied by a parent, set off for new waters. Migrating male Razorbills and their chicks, for example, swim—the chick while it grows and the adult still flightless from its molt.

Flight power and speed

Calculating the optimum speed to cover the maximum distance requires expert tuning.

The power of flight depends on a number of complex factors that birds must integrate for maximum effect. Obviously, they do not consciously work out a series of mathematical equations, any more than a baseball catcher or fielder uses spherical trigonometry to reach and catch a ball. Nonetheless, migrants must be aware of these factors, since mistakes can be a matter of life and death. A bird that sets off too fast on its journey may use its energy too quickly and run out of fuel before it reaches its destination. If it is crossing a desert, this may be fatal.

For a bird in flapping flight, two main activities consume power, each of which varies with the speed of the bird. The first—the power needed to keep the bird aloft—decreases the faster the bird is flying. Some birds are able to hover for a short length of time, and hummingbirds and sunbirds can even fly backward, but for most birds energy consumption is so high at low speeds that they are unable to fly slowly for any length of time. The second factor—the drag induced by the airflow over the bird's wings—increases power consumption the faster the bird flies.

Although the combined effect of these two forces differs according to the bird's size and shape, for many it means that there is both a speed below which they cannot maintain flight and one above which they cannot go. In between is a fairly wide range of maintainable speeds, one of which can be sustained with a minimum power input. This speed is the one that will provide the maximum time that the bird can fly without stopping to "refuel."

The speed that requires the minimum power input and produces the maximum time in the air is not necessarily the bird's maximum range speed, the speed at which it achieves its maximum range (in theory, the greatest distance it can cover without refueling and at minimum power). This is because maximum range also depends partly on the power the bird needs to stay alive—in other words, its metabolic energy—and on circumstances. If an extra 5 percent of speed needs only 3 percent more energy, the bird will achieve its maximum range in a shorter time and at a higher speed.

Such parameters are under the bird's control; the wind, by contrast, is not, yet it may have a huge effect on the bird's speed, energy consumption, and range. At worst, if the wind speed against the bird is greater than its maximum speed, it can make no headway. If the wind is in the bird's favor, the distance traveled over the ground at its maximum range speed may be greatly enhanced.

FLIGHT AND WING SHAPE

A migrant's wing shape and size are related to its flight. Large sea birds (top) have long thin wings for dynamic soaring low over the waves. Broad, fingered hawk wings are best for soaring over land, allowing fine control of the airflow at their tips. Large birds such as geese have wings that are heavy for their size for flapping and gliding. The short, rounded wings of small birds (bottom) are perfect for flapping flight.

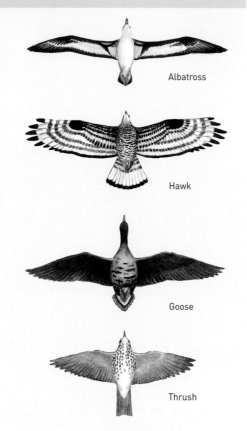

Albatross

Hawk

Goose

Thrush

POWER SPEED CURVE

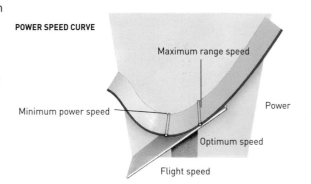

Maximum range speed

Minimum power speed

Power

Optimum speed

Flight speed

THE GRAPH (above) shows the relationship between power consumption and speed of flight. Power consumption is high until the bird reaches its speed of minimum power input and high again once it has passed its maximum range speed. For this reason, although actual speeds vary from bird to bird, most flapping birds fly at a speed that falls between these two points (pink shading).

The wing of a Cetti's Warbler is short and broad with short primary flight feathers

Cetti's Warbler Asian race *albiventris*

The wing of a Reed Warbler is longer and narrower with long primary flight feathers

Reed Warbler

CETTI'S AND REED WARBLERS frequent similar habitats: in fact, in places their ranges overlap and they share territories. Their wings, however, are completely different: Cetti's Warblers' are short and broad, Reed Warblers' are long and rather narrow. Their migratory habits also differ. Reed Warblers are long-distance migrants while Cetti's Warblers are resident. The constraints that are imposed by this difference determine their wing shapes.

Short, broad wings are ideal for a life spent in thick, tangled vegetation foraging for small, agile insects. For efficient long-distance migration, wings must be long and thin to ensure energy-efficient flight.

Soaring birds are affected by different constraints from birds that flap to fly. Relatively little power is used to keep wings outstretched and to make any necessary small adjustments to the control surfaces. Measurements suggest that the power saving over flapping flight may be as great as 95 to 97 percent. This means that many soaring birds of prey are able to make their journeys relatively slowly and without feeding, an important factor for birds that are funneled through highly specific areas. The hundreds of thousands of hawks and buzzards that use the Isthmus of Panama or the Strait of Gibraltar on migration have little chance of competing with each other or with native birds for local food stocks. Instead, they fly when the thermals are working well and at other times roost to conserve energy.

At sea, the situation is similar. Species using dynamic soaring to travel show power savings of the same order of magnitude as land-based soaring birds. A newly fledged Manx Shearwater —which weighs about the same as a healthy adult—sets out immediately from the island of its birth to the wintering grounds of Brazil. Since the intervening 6,000 miles (10,000 km) of ocean offer little chance for surface-feeding birds to find food, the young shearwater probably does not feed en route. The fastest passage time so far recorded is 17 days, but birds' reliance on the wind can cause problems: satellite-tracked Wandering Albatrosses have been "grounded" when the winds are not strong enough for dynamic soaring (see p. 45).

Large amounts of fuel have an effect on the aerodynamics of a bird's flight. The way that the fat is laid down alters the bird's profile (see p. 24), and it must expend more energy to carry the extra weight. Calculation of maximum range, therefore, may be different at the beginning of the migration when a bird is heavy with fat, than at the end, when it has consumed up to half its start weight.

Air density also affects range calculations. Density is reduced at greater heights, which makes it harder for a bird to find lift at a given speed, but for the same speed, drag is reduced.

How high do birds fly?

Flying high brings death to migrants in collisions with airplanes.

Birds reach their height limit when the reduced amount of oxygen in the air, and the air's lower density, prevents them from functioning normally. For some species this is at a great height: Bar-headed Geese have been recorded above the summit of Everest. Individual migrant birds, however, fly at the height that makes most sense on the journey they are undertaking at the time.

The most important determining factor is the height above sea level of the ground below the migrant. Birds negotiating the Alps via Col de Bretolet must fly at an altitude of some 6,600 feet (2,000 m) to get over the pass. If they are flying against the wind, they may be only 10–13 feet (3–4 m) above the ground along the pass.

Another constraint on height is the presence of cloud. Birds prefer not to fly in cloud, and heavy cloud greatly impairs their performance. This is because birds in flight produce so much water when they burn fat in their muscles as fuel that getting rid of it in cool, damp air is difficult.

Wind speed and direction

The wind is a vital consideration for most migrants. Close to the ground, the wind is less strong than it is higher up—certainly above 1,600 feet (500 m) from the ground. Higher still, wind speeds and directions may be completely different from those near the ground—the wind an observer feels on the ground is often blowing in a different direction from the wind that

Feet

Whooper Swan

25,000

20,000   Bar-tailed Godwit

15,000

Fieldfare

10,000

Swift

5,000

Chaffinch

feet   0

HEAD AND TAIL WINDS

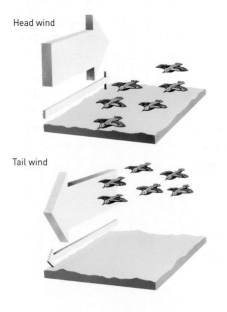

Head wind

Tail wind

For the migrating bird, wind is the environmental factor that has the greatest effect. Many small birds have a flight speed that gives them a maximum range for each unit of energy consumed of some 25 mph (40 km/h), so that even a relatively light direct head wind of 12 mph (20 km/h) reduces their range by 50 percent, and a gale of 30 mph (50 km/h) would blow them backward. Birds have no problem in deciding the strategy for a moderately light head wind—stay low and hug the ground. In a tail wind, the best option is to get up high and fly with it.

SEVERAL METHODS are used to measure the height at which birds fly (right). Radar gives accurate height readings and allows a degree of species identification by measuring size, speed, and wingbeat from the returning echo. The remains of birds hit by aircraft are logged, particularly if the birds are large enough to put aircraft at risk.

The heights shown in the illustration are the maximum recorded. Since the altitude of flight depends on wind, cloud, and height above sea level, these heights may not be regularly attained.

Bar-headed Goose

Mallard

White Stork

Lapwing

Tundra Swan

Snow Goose

Black-bellied Plover

Black-and-white Warbler

HIGH FLIERS include Bar-headed Geese, which cross the Himalayas at heights of up to 29,500 feet (9,000 m) as they travel between the mountain lakes of central Asia and their winter homes along the Indus valley, India. A flock of 30 Whooper Swans en route from Iceland to western Europe was logged by a pilot at 27,000 feet (8,230 m). Mallards have reached 21,000 feet (6,400 m), Bar-tailed Godwits 19,685 feet (6,000 m) and White Storks 15,750 feet (4,800 m) on migration.

Flocks of Lapwings often travel at modest altitudes, but have been sighted at 12,800 feet (3,900 m), while Fieldfares may reach 10,800 feet (3,300 m). Whistling Swans may fly at 8,850 feet (2,700 m) across North America. Swifts, among the most aerial of all birds, attain heights of 6,600 feet (2,000 m); Snow Geese have been noted at 4,900 feet (1,500 m); and Black-bellied Plovers at no more than 2,600 feet (800 m).

Many small birds keep relatively close to the ground. Among them, Chaffinches are usually well below their maximum of 3,300 feet (1,000 m) while the Black-and-white Warbler and other wood warblers reach a maximum altitude of around 1,600 feet (500 m).

is sending clouds racing across the sky. For this reason, if birds have to fly against the wind, they are likely to be low and clearly visible, an excellent spectacle for birdwatchers. On another day, there may be more migrants, but if the wind is behind them, they are almost certain to be flying so high that they are out of sight.

In a cross wind, the birds must take advantage of the wind speed and strength without being blown off course. The way to do this is to aim to the right (if the wind is blowing across them from the right) or the left of the ultimate target. By deliberately allowing for the effects of the wind in this way, the birds use it efficiently and reach their destination with no need for corrections at the end of the journey.

Radar observations often show migrants shifting their altitude to try to find the best height with the greatest amount of wind in their favor. It is assumed that they are able to gauge how the wind is affecting them by measuring their drift against the distant horizon.

Preparing for the journey

Fat, fit, and with new feathers, birds are ready to undertake their migratory journeys.

All sorts of preparations are necessary before a migrant bird can set out on its journey. For most migratory species the late summer is the time for laying down fat reserves to use as fuel on the journey, changing their feathers, and improving the power of their flight muscles to make sure that flight is swift and efficient.

The most obvious preparation for migration is to put on subcutaneous fat, just beneath the skin. All species have to do this, unless they are able to feed while migrating—and even then, a bird may need substantial reserves if its journey involves desert or sea crossings. British Garden Warblers, for example, which usually weigh about ½ ounce (12 g), put on around ⅐ ounce (4 g) of fat to take them across the Bay of Biscay to northwestern Iberia. There, they feed intensively again, increasing their weight by about ⅓ ounce (10 g) so that they can fly over the rest of Iberia, northern Africa, and the Sahara in an epic three-day journey. Immediately south of the Sahara, a further small increase in fat provides fuel to take them to their wooded savanna wintering areas.

Spring weight increases in Africa are even more impressive as the journey across the desert is likely to be against head winds. Increases of 100 percent on "normal" weight have been recorded, which would allow a flight of 2,500 miles (4,000 km) with no wind, over up to five days.

Such weight increases are accomplished through changes in behavior that are triggered by hormonal activity. These have other effects on the bird, again setting in motion changes that are necessary before the journey can start. Like an athlete training

White-fronted Goose (migratory) Molt takes approx. 3½ weeks

Chaffinch (resident) Molt takes approx. 10–11 weeks

Chaffinch (migratory) Molt takes approx. 8 weeks

Snow Bunting (migratory) Molt takes approx. 4 weeks

PRIOR TO MIGRATION a bird lays down fat both within its body cavity and underneath its skin. This is easy to see on a bird in the hand. Blowing the feathers upward reveals the transparent skin: the pale fat contrasts with the pink or red flesh. The best place to look for fat on small birds is the furculum, the pit above the breastbone.

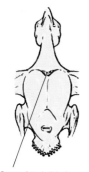

Some fat visible in furcular region

Furcular region almost filled with fat

Furcular region filled with fat

THE ANNUAL MOLT of the main flight feathers in most species takes place gradually, so that the birds can continue to fly. The primary feathers (those nearest the wingtips) molt from the inside outward. Then, when the inner ones have regrown, the bird starts to shed its secondaries from the outside inward.

The speed of the molt (above) depends on circumstances. Resident Chaffinches, for example, may take 10 to 11 weeks to complete their feather change; migratory Chaffinches, on the other hand, take less than eight weeks. The fastest perching birds—at four weeks—are probably Greenland-nesting Snow

Buntings. These birds are flightless for a time.

Waterfowl and other groups become totally flightless during the molt and replace all their flight feathers at once—White-fronted Geese cannot fly for some 25 days. (Timing of the molt differs between populations, but is of the duration specified during the months indicated.)

MOLT MIGRATION

Some species, particularly waterfowl, undertake special migrations for the purpose of molting. These seem to be governed by the presence of food stocks that cannot be exploited by breeding birds: often they are available only late in the breeding season.

Sometimes the birds that leave the breeding area are failed breeders and nonbreeders which go, perhaps, to avoid competition for resources with growing families. Often, however, molt migrants are successful parents that leave their young in the care of "aunts," surrogates that look after "nurseries" of young from several broods.

This may be the case with the Goosanders, that molt in Finnmark, Norway, and is the strategy used by Eiders (right) and Shelducks. Shelducks have several molting grounds in Europe, the best known of which is the Knechtsand, Germany, where up to 100,000 birds gather each fall.

for a marathon, a bird needs to tone up its whole physique, so its major flight muscles increase in size. In an average bird, these muscles account for some 15 percent of its total weight, perhaps $1/14$–$1/12$ ounce (2–2.5 g) in a medium-sized species. When that bird is fully fat prior to migration, the amount of its body weight taken up by the flight muscles may be reduced to 12 percent, but the weight of the muscles themselves has probably risen to $1/10$ ounce (3 g).

In some circumstances—if a bird is held up by contrary winds, for example, and runs out of fuel on migration—it can convert some of this extra muscle tissue into fuel to enable it to continue its journey. It is still not certain whether this happens as a matter of course (as the additional muscle was to enable the bird to

cope with the extra weight of fat, it is not strictly necessary) or whether it is something that occurs only in emergencies.

Finally, many birds prepare for one or another of their seasonal migrations by molting their flight feathers. The molt is also hormonally controlled and essential for the birds because their feathers suffer wear and tear. Since the efficiency of flight depends on the quality of a bird's wing feathers, their renewal immediately before migration is sensible. And, at this time of the year, food is plentiful so the birds are sure to have the energy necessary to complete a successful feather replacement. Young birds do not need to renew their feathers prior to their first fall migration, but may do so on the wintering grounds before the return flight in spring.

Timing

Deciding when to leave on migration can be a tricky business.

Migration is essentially about being in the right place at the right time, so in the broadest sense timing is crucial. The means by which migrant birds navigate (see pp. 30–33) also make precise timing, on a daily basis, necessary. For these reasons, birds have two "body clocks" built into their physiology, to give them good circannual (yearly) and circadian (daily) rhythms. Birds are not alone in having these two clocks—many organisms, including bacteria and humans, share them—but they are obviously of paramount importance for migrants.

The presence of circannual rhythms has been proved by keeping caged birds under a constant light/dark cycle, usually 12 hours of each. In these conditions, birds of several different species breed, molt, show migratory restlessness (Zugunruhe), and put on weight in a normal yearly cycle. If the year is compressed by accelerating the natural changes in day length, the birds speed up their functions: the record is held by a starling that was tricked into undergoing eight annual molt cycles in one year.

The change in the light/dark cycle is most striking at the equinoxes in temperate latitudes, when the rate of change from dark to light is particularly apparent. It is not so obvious in the tropics at any time, or in the Arctic in high summer, so while this environmental stimulus is important to the circannual rhythm, it is not the only factor involved. Circannual rhythms are probably not changed by periods of rain, abnormal temperatures, or the growth of vegetation, but the birds' behavior is modified by such factors. In a warm spring, they may breed a week or 10 days earlier than in a year when temperatures are "normal."

Shortening days and a bird's internal clock act on its hormonal system, stimulating the physiological changes associated with fattening (see pp. 24–25) and the molt. Later, the bird's hormonal balance, and probably also its weight, stimulate migratory restlessness. In the wild, a bird at this stage would migrate; in a cage, it makes restless movements oriented toward the direction in which, if uncaged, it would be moving.

Birds are aware of the weather and its effect on the success or failure of a migratory journey. They do not generally migrate when the sky is heavily overcast, in strong contrary winds, or in rain. They are often, however, able to predict favorable winds from prevailing air conditions. The result is that birds migrate in waves—held up by bad weather, many birds move on together as soon as conditions improve. Such considerations determine the precise day or night the birds begin to migrate, so that while the week of their movement is generally preordained, the date is determined by environmental factors.

The circadian rhythm is a different matter. Under continuous dim light, caged birds are as likely to settle down to a day length of 23 or 25 hours as 24. In real life, environmental clues, or Zeitgebers, keep the bird on a strict 24-hour cycle. The most important factor by far is the daily light/dark cycle, but in constant dim light, a 24-hour temperature cycle that mimics the external world keeps caged birds on a 24-hour cycle. In many species, but possibly not all, the action of this clock is governed by the pineal gland, situated in the front of the brain.

For use in true navigation, the circadian rhythm would have to be staggeringly accurate. The sun is at its highest in the sky at noon: an error of 10 minutes in judgment of time would put a migrant 125 miles (200 km) off target at the latitude of New York or Paris. It is more likely that birds use the sun as a directional aid: navigating with a time-compensated sun compass gives rise to an error of 12 miles (20 km) after a journey of 300 miles (500 km).

The pressures on birds to migrate particularly early are strongest for the males of most species returning to the breeding grounds in spring. An early return gives them the opportunity to occupy the best territory and thereby mate with the best females. If they arrive too early, they may find poor weather and risk their survival, although all is not necessarily lost. Early birds, so long as they still have sufficient fat reserves, may be able to retrace their steps and try again later. In some of the larger species, these early migrants seem to act as scouts. If they set out and encounter poor conditions, such as the ice not having broken, they return to where they started. The birds left behind then assume there is no point in their trying yet. If the scouts do not come back, they realize that they should perhaps try the journey for themselves.

A MIGRANT DESERT WHEATEAR (right) experiences a variety of stimuli. Its internal clock may be telling it to go, but it is also sensing what the weather is doing. Migrating into a storm is not a good idea, but the storm may be part of a weather system that will leave favorable tail winds in its wake. The Wheatear cannot take advantage of this opportunity if it has not put enough fat into its internal stores for the next leg of its journey. In this case, the need to provision itself is paramount, and the bird must look for the best place to find food.

Wheatears on migration almost always leave at dusk since this is when they can see the setting sun and establish where west is. After that, they rely on the stars for an accurate directional compass.

Genetics and migration

Most young birds know instinctively which way to go and how to find the way back.

The true wonder of migration is that it is instinctive—generation after generation, birds are able to undertake these often complicated patterns of movement. It is obvious that young cuckoos, left by the parents that migrate early, must be able to find the way alone. It has been shown that many other birds' genetic makeup is also imprinted with enough information to enable naive juveniles to accomplish their regular migrations without the benefit of any help from their parents.

Indeed, in the majority of species, there is little chance of the young remaining in touch with their parents. In preparation for their first return flight the following spring, many juveniles spend some time exploring their home area before they depart on their first migration. This period coincides in many species with the time that their parents are raising a second brood, inevitably a full-time occupation, or undergoing their full molt.

The juveniles' exploration includes local features such as the adjacent territories, but also takes several species some 30–60 miles (50–100 km) from their nest sites. The regular appearances in the fall in western Europe of eastern European birds that migrate southeastward, such as Barred Warblers and Red-backed Shrikes, may be part of this familiarization process. It is clear that other species that would normally be found in western Europe at this time—but are absent—also venture away from their birthplaces in this way. Blackcaps from the Scandinavian population, which migrate southeastward, have been found in Britain and then picked up again in Turkey.

As if the "program" for accomplishment of the journey were not enough, birds also inherit the annual rhythm that triggers the necessary physiological changes, so that they are ready to migrate at the right time. Equally important, the rhythm stops the cycle when they reach their winter quarters (see pp. 24–27).

The genetic basis of migration causes birds to be conservative in their choice of wintering area. It is no accident that Willow Warblers from all over Eurasia, for example, still winter in Africa (see also Chats and thrushes, pp. 108–9). The location was "encoded" in their ancestral genetic material, and no mutation has happened to provide a population with a viable alternative. Even during the warmest interglacial of the last cycle of ice ages, which ended 10,000 years ago, the wintering area did not shift far enough north to prompt a spread to Asia when the ice returned. Gradual change is always possible—sudden and radical change is not.

AN EXPERIMENT with Common Starlings (below) proved the roles of experience and genetics in migration. Birds caught on the fall journey in the Netherlands were taken to Switzerland and released. The juveniles, which had not migrated before, continued in their "programmed" flight direction and reached Spain. The more experienced adults corrected course to reach their winter homes in France, Britain, and Ireland.

WHOOPER SWANS (right) migrate in family groups so that the young benefit from the adults' knowledge of previous migrations. The young could almost certainly manage alone, but they would be at a disadvantage, lacking site-specific details of the wintering areas that are subject to short-term change.

Juvenile

Adult

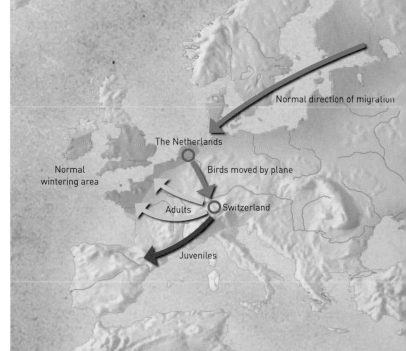

Normal direction of migration

The Netherlands

Normal wintering area

Birds moved by plane

Adults

Switzerland

Juveniles

AS WITH MANY MIGRANTS, Blackcaps (below) are divided in their migratory paths. Birds breeding in western Europe fly to the southwest, while those from eastern Europe move southeastward.

An experiment at Radolfzell in southern Germany crossed birds from both populations. As expected, the hybrids tried to fly south—and would have failed to survive on a path that took them over the Alps and the widest part of the Mediterranean. Hybrids that occur in the wild also die; survival lies in keeping to the migratory divide.

The changing nature of Blackcap migration illustrates this gradual process, as does the manner in which Evening Grosbeak migration has altered in this century (see p. 13). British Blackcaps are migratory, heading southward to winter from southern France south to the Sahel. There are, however, records of increasing numbers of Blackcaps wintering in Britain, in locations as far north as Inverness, Scotland, and in Ireland. These are German and Austrian migrants. During the winter, up to January, they feed on berries, but during the first three months of the year, they more frequently take the offerings from bird feeders.

Blackcaps are not strong migrants, but they normally travel southwestward. Since their direction of travel is not exact, a few almost certainly reached Scotland and Ireland naturally. Conditions, helped by the numbers of well-stocked bird feeders, favored their survival, and natural selection has increased the numbers taking this course.

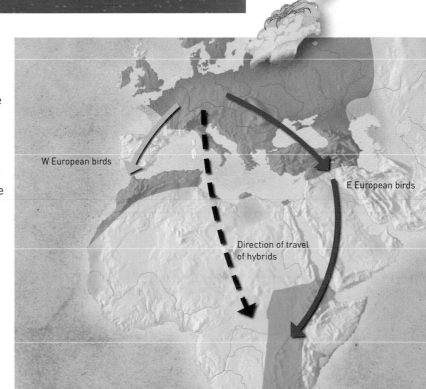

W European birds

E European birds

Direction of travel of hybrids

Orientation and navigation ①

Getting there involves flying in the right direction for the correct length of time.

Navigation is the art of getting there. This generally means that a migrating bird has to know the direction in which it must fly— it must be able to orient itself—and to know when it has arrived, both of which are easier if the migrant has done it all before. For many years it seemed that bird navigation would never be understood, since observers were looking for a single answer to the question "How do birds find their way?" It is now known that there is no one answer but that birds use a host of clues. It is akin to a walker being lost: he or she might use a watch and the sun to find south, locate the moss beside a tree to detect the damper west side, walk down a river bank to reach the sea, or wait until dark when the North Star appears. A bird, similarly, cannot afford to ignore any clues.

For many birds, simply flying in the right direction for the correct length of time should be good enough to get from one place to another. Certainly, the duration of migratory activity in caged Blackcaps is related to the distance that the different populations to which they belong are moving in the wild. This, together with the recognition of suitable habitat, is enough for a small bird with a fairly extensive wintering area to accomplish its "outward" migration.

The return journey is more crucial, but here, too, birds are looking for a large area with which they are familiar and, after they have bred, they almost certainly have a real target to aim for. Most birds have the ability to navigate to within about 6 to 12 miles (10 to 20 km) of their goal, after which they use landmarks to reach their precise destination. "Bird's-eye view" also helps— landscape features that are invisible to ground-based observers can be seen from above.

Sun, stars, and magnets

There are three main means—all compasses—by which birds detect absolute north, south, east, and west, namely the magnetic, star, and sun compasses. The most difficult of these to understand is probably the magnetic compass, but it is likely that the ability to locate magnetic north depends on small crystals of magnetite situated above the nostrils. An experiment using students demonstrated that humans have a similar sense of direction.

THE EYES OF A MIGRANT BIRD receive major navigational clues from the stars at night and the sun by day. They also allow it to note landmarks so that the bird can orient itself and recognize specific features once it has returned to a familiar area.

A bird uses its sense of balance, together with what it can see of the land surface, to assess the effect of the wind. This may be through drift from its preferred direction, or from lift. Its ears may hear infrasound from the wind on distant mountains or the soft contact calls of birds nearby. Smell or taste may be used to locate the pungent colonies of sea birds—or the sweet scent of meadows.

Within the bird's head, its innate sense of direction also has an effect on its brain. And all these stimuli are modified by the bird's internal clock.

Stars

Moon

Magnetic north

Polarized light

Sunrise

To destination

Weather fronts

UV light

Winds

Landscape features

Sound waves

Smells

All the students wore coils on their heads, some of which canceled out the earth's magnetic field, while the rest left it unaffected. At the end of a tortuous bus ride, only those wearing the "dummy" coils could point toward home.

European Robins tested for a sense of direction under changed magnetic fields show that they are able to orient well—but they have to be able to move around, crossing the lines of the magnetic field, to do so. Magnetic north itself moves around, and although it may provide a good reference point for a bird in its lifetime, over a few thousand years it may shift by up to 30°. Birds are able to cope with the regular reversals of polarity which happen every few hundred thousand years. It is believed that this change takes several years to stabilize; during this time, any species relying solely on its magnetic sense would be unable to complete its migrations. Birds, therefore, must be able to make use of other means of navigation.

SUN COMPASS

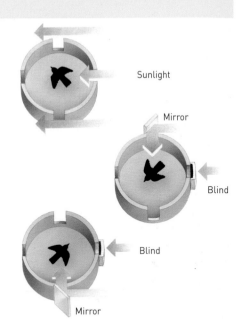

Experiments to prove the existence of a sun compass and its importance for daytime migrants involved caged birds and mirrors. At migration time, in natural conditions, these birds clearly showed a preferred flight direction (top).

When mirrors bent the sun's rays through 90°, the birds turned their preferred direction of movement through a right angle (middle). When the sun's rays were modified by the same amount in the opposite direction, this became the focus of the birds' interest (bottom).

Sunlight

Mirror

Blind

Blind

Mirror

MAGNETIC COMPASS

Lines of magnetic force

Robins were placed in cages surrounded by magnetic coils that mimicked the earth's magnetic field (visualized, top).

Previous work had shown that these birds register the angle made by the field and the surface of the earth—which points south in the northern hemisphere and north in the southern.

Here, they detected north, the direction of their spring migration ❶. When the field was twisted so that north was in the east-southeast, the birds kept to their original path for the first two nights ❷. By the third night, they had detected and taken account of the change ❸.

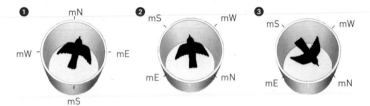

❶ mN mW mE mS

❷ mS mW mE mN

❸ mS mW mE mN

STAR COMPASS

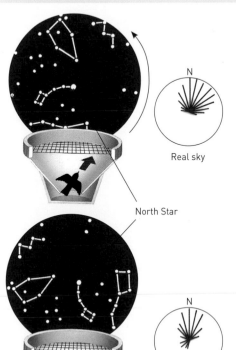

North Star

N

Real sky

Here, birds stood on an ink-pad at the base of a cone of blotting paper. Nocturnal migrants flutter in their preferred direction of travel: the amount of ink shows which that is. In a planetarium projecting the real sky, Indigo Buntings located the North Star and used it to find north, the direction of their spring migration (top). When the sky was rotated, the birds still oriented northward (bottom). When the sky was obscured, as on cloudy nights, they became confused (inset).

N

Sky rotated 90°

N

Sky obscured

Orientation and navigation ②

Tests suggest that birds take some time to perfect their ability to use the magnetic compass effectively, and in some cases, they appear to use it to calibrate their star maps. Birds' use of the star map itself is now well understood as a result of a series of experiments in planetaria. Birds are not hatched with a built-in star map in their memory, but they do have the ability to detect the center of rotation of the stars. At the moment, this is near the North Star, but because the earth's axis precesses, it does not stay there: in 13,000 years' time, it will have moved by 47°. Clearly this is enough to make a huge difference for a migrating bird, and the star map would not work effectively; detecting the center of rotation, on the other hand, is perfectly safe.

The sun compass relies on the bird's internal clock (see also pp. 26–27). At 10 a.m. local time (which a bird detects with its "body clock"), the sun is always 30°E of south. Several species make use of the position of the sun, relative to the time of day, as a directional guide. Others may use the fact that it rises in the east and sets in the west to aid navigation. It has also been suggested that an internal feature of the eye—*the pecten oculi*—may, like the gnomon of a sundial, throw a shadow on the back of the eye to provide special clues on direction. A sight of the sun is not necessary for its use in navigation—light from the sun coming through the clouds is polarized, which the birds can detect.

Sensing the way

It has become clear that the sense of smell may be used by some birds during movements—this has been proved in experiments with homing pigeons and is suspected for many petrels, which have strong and characteristic scents.

Birds may use sounds in various ways. The calls of amphibians in marshy areas, and of other birds, waves on shores, and infrasound caused by the wind on mountain ranges have all been thought to offer positional clues. It is even possible that birds can use the echoes of their own calls from the land surface in this way.

Migrants crossing the oceans can probably detect the presence of land by looking at the overall pattern of the waves. Barometric pressure may provide useful clues—but more probably to help in making the journey efficiently rather than in navigation. Recent results also show that birds may be conscious of small changes in gravity and are possibly able to detect the Coriolis force—the force produced by the spinning earth.

MIGRATION GENERALLY takes place on a broad front although many natural features funnel or concentrate migrants. Such features—known as leading lines—may, in addition, provide them with clues to navigation. Many leading lines are also excellent places to observe migrants.

Mountain ranges (right) can be huge obstacles, so migrants tend to be funneled through passes. Small islands and headlands are also points where migrants congregate.

Low mountains and hills can provide ideal soaring conditions for specialist migrants, and coastal areas may concentrate birds relying on thermals for lift. Such birds also make use of the narrow joins—such as Panama in Central America—or narrow seaways—such as the Bosporus in Turkey—between landmasses. Land birds also use valleys and sea or lake coastlines pointing in roughly the right direction as leading lines.

Routes and barriers

No barrier seems insurmountable when the migration route is generations old.

Migration routes, built up over generations, cannot change overnight. The genetic basis of migratory control (see pp. 28–29) makes change possible only through gradual evolution. This process may take place over a long period and cover huge distances—Northern Wheatear migrations change by about 5/8 mile (1 km) a year, but have been doing so for almost 10,000 years. Routes, of course, are only sustained if birds are able to use them efficiently and if enough birds survive to maintain the population.

A good route between summer and winter quarters need not be the most direct. There is often a barrier to migration that it is prudent for the birds to skirt. This may be sea for land birds, or land for sea birds, or inhospitable ground such as desert or a mountain range. Such obstacles do not necessarily hinder the birds, which may decide to overfly the barrier even if there is no chance of landing to refuel.

Birds that rely on special flight techniques have no choice in such matters. Thermal soarers cannot fly at night, over the sea, or in any area where they cannot achieve lift. And parts of the ocean are no-go areas for sea birds—when the winds are unreliable, dynamic soaring is not an option.

All sorts of barriers have shaped and influenced migration routes. The watery expanse of the Caribbean and Gulf of Mexico in the Americas, the vast Sahara desert in Africa, and the high peaks of the Himalayas in Asia are all effective barriers to migration. But none is absolute—many species have evolved routes across them.

GREAT CIRCLE ROUTES

The shortest distance between two points is rarely apparent from conventional maps, which flatten the earth's sphere. Birds, like jet airliners, take routes that follow the circumference of the earth, so-called Great Circle routes (below, left). These result in considerable distance savings, particularly for long-distance migrants.

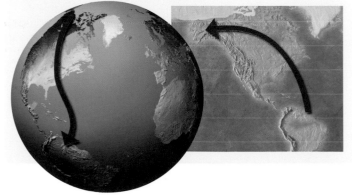

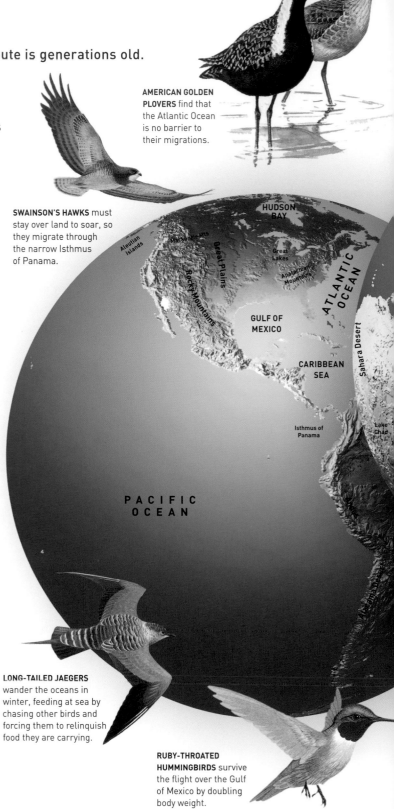

AMERICAN GOLDEN PLOVERS find that the Atlantic Ocean is no barrier to their migrations.

SWAINSON'S HAWKS must stay over land to soar, so they migrate through the narrow Isthmus of Panama.

HUDSON BAY

Aleutian Islands

Mackenzie Mts

Great Lakes

Great Plains

Rocky Mountains

Appalachian Mountains

ATLANTIC OCEAN

GULF OF MEXICO

Sahara Desert

CARIBBEAN SEA

Isthmus of Panama

Lake Chad

PACIFIC OCEAN

LONG-TAILED JAEGERS wander the oceans in winter, feeding at sea by chasing other birds and forcing them to relinquish food they are carrying.

RUBY-THROATED HUMMINGBIRDS survive the flight over the Gulf of Mexico by doubling body weight.

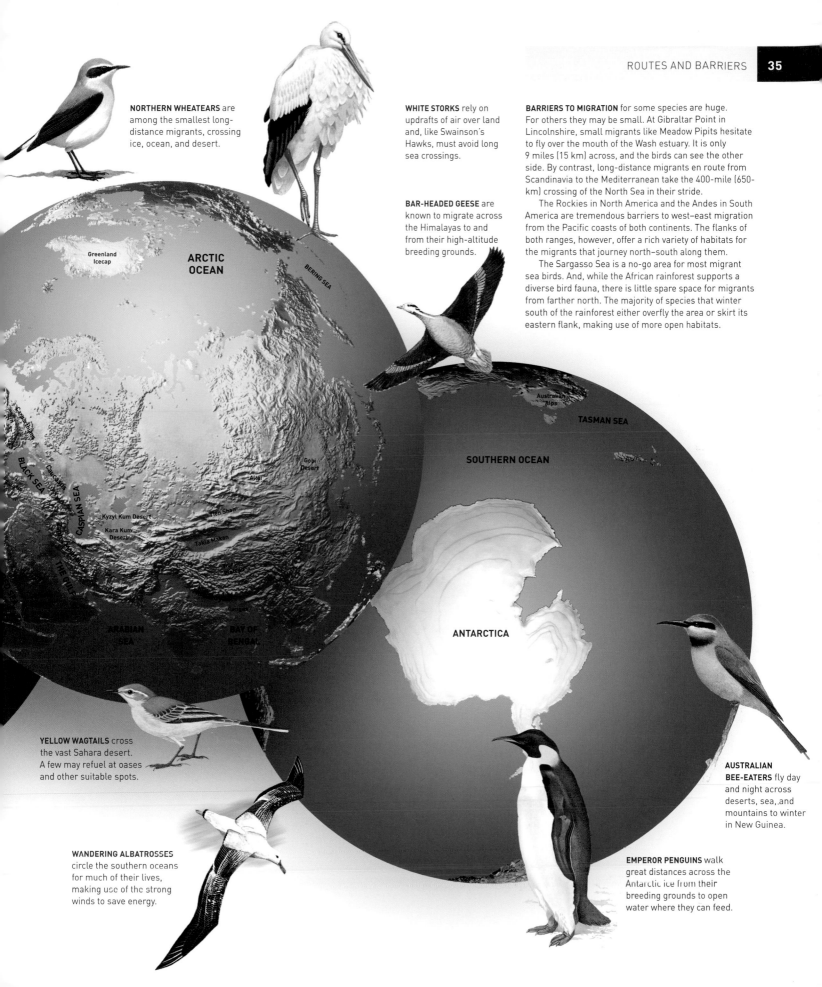

NORTHERN WHEATEARS are among the smallest long-distance migrants, crossing ice, ocean, and desert.

WHITE STORKS rely on updrafts of air over land and, like Swainson's Hawks, must avoid long sea crossings.

BAR-HEADED GEESE are known to migrate across the Himalayas to and from their high-altitude breeding grounds.

BARRIERS TO MIGRATION for some species are huge. For others they may be small. At Gibraltar Point in Lincolnshire, small migrants like Meadow Pipits hesitate to fly over the mouth of the Wash estuary. It is only 9 miles (15 km) across, and the birds can see the other side. By contrast, long-distance migrants en route from Scandinavia to the Mediterranean take the 400-mile (650-km) crossing of the North Sea in their stride.

The Rockies in North America and the Andes in South America are tremendous barriers to west–east migration from the Pacific coasts of both continents. The flanks of both ranges, however, offer a rich variety of habitats for the migrants that journey north–south along them.

The Sargasso Sea is a no-go area for most migrant sea birds. And, while the African rainforest supports a diverse bird fauna, there is little spare space for migrants from farther north. The majority of species that winter south of the rainforest either overfly the area or skirt its eastern flank, making use of more open habitats.

Greenland Icecap

ARCTIC OCEAN

BERING SEA

CASPIAN SEA

BLACK SEA

Caucasus

Kyzyl Kum Desert

Kara Kum Desert

Takla Makan

Tien Shan

Altai

Gobi Desert

THE GULF

ARABIAN SEA

BAY OF BENGAL

Australian Alps

TASMAN SEA

SOUTHERN OCEAN

ANTARCTICA

YELLOW WAGTAILS cross the vast Sahara desert. A few may refuel at oases and other suitable spots.

WANDERING ALBATROSSES circle the southern oceans for much of their lives, making use of the strong winds to save energy.

EMPEROR PENGUINS walk great distances across the Antarctic ice from their breeding grounds to open water where they can feed.

AUSTRALIAN BEE-EATERS fly day and night across deserts, sea, and mountains to winter in New Guinea.

Staging posts

Ensuring a successful migration often means stopping off en route.

Birds that complete their migratory journeys nonstop are the exception rather than the rule. Most migrants have to stop over somewhere to rest and replenish their fuel reserves. For many small birds, these stopping places are extensive and the birds not particularly concentrated or apparent to observers, but some of the larger species gather in spectacular numbers in clearly defined areas.

The main function of a staging post is to provide a place where food is plentiful so that birds can fatten up for the next stage of their journey in safety. The location of a migration stopover for a species may, therefore, differ in spring and in fall. Similarly, the length of time a bird spends at a staging post varies from species to species. Depending on the availability of food and the speed with which a bird can put on fat, it can be as little as a few days or as much as a few weeks.

Staging areas where large numbers of migrants become concentrated are indicated on the map below. Among the smaller migrants using more diffuse and less easily defined stopovers are Reed Warblers from western Europe, which become concentrated down the Portuguese coast in August and September; Blackpoll Warblers from much of eastern Canada, which spend time in Massachusetts in the fall; and Pied Flycatchers from across western Europe into Asia, which are found in northwestern Iberia in the fall.

○ **THE STAGING POST SYMBOL** can be found on many of the maps in the chapters that follow to indicate a traditional area where migrants concentrate to feed and rest in spring or fall.

1. Aleutian Islands
2. Copper River delta
3. Gray's Harbor
4. Tule Lake
5. San Francisco
6. Freezeout Lake
7. Devil's Lake
8. James Bay
9. Platte River
10. Cheyenne Bottoms
11. Galveston/Live Oak/Aransas NWR
12. Cameron
13. Dauphin Island
14. Lake Erie
15. Delaware Bay
16. Bay of Fundy
17. Cuba, and other Caribbean islands
18. Panama
19. Paracas
20. Surinam coast
21. La Serena, Chile
22. Tierra del Fuego
23. Lagoa do Peixe
24. Bear Island
25. Ammassalik
26. Iceland
27. Bals fjord
28. L. Hornborgasjön
29. Gulf of Matsalu
30. Morecambe Bay/ The Wash
31. Waddenzee/IJsselmeer
32. Rugen
33. Caspian Sea
34. Banc d'Arguin
35. Lake Chad
36. Benguela Current
37. Islands of South China Sea
38. Coasts of Western Australia
39. Port Phillip Bay/ Corner Inlet

SOME ESTUARIES are used both by birds on passage and as wintering grounds, with the result that they are important year-round. Delaware Bay (left), on the east coast of the U.S., is a special case. At the end of May, vast numbers of horseshoe crabs lay their eggs here, and every year Red Knots of the race *rufa* come to feast on this predictable food stock. At least 60,000, and perhaps as many as 95,000, birds used to gather here, more than half the world population of this race, However, in recent years, over-fishing of shorecrabs has led to a great reduction in Red Knot numbers.

The birds pack close together, unconcerned that a neighbor might be in competition—there is plenty for all.

✪ WATCHING THE BIRDS

Some staging posts are superb places to see large numbers of migrant birds, others less so. Other good sites for birdwatchers are places such as Hawk Mountain, Pennsylvania, where soaring birds are concentrated by geographical features. Similarly, at Point Reyes, California, and Fair Isle, Scotland, migrants concentrate due to quirks of geography.

The star symbol recurs throughout the book at sites where migrant birds, for whatever reason, are particularly visible. These may or may not also be staging posts. (See Watching the shorebirds, pp. 126–27.)

Weather and climate ①

Correctly assessing weather is vital, but mistakes may be rectified.

On a day-to-day basis, the weather can be of overriding importance for a migrant. For individual populations and for individual birds, all aspects of the weather—warmth, cold, dryness, rain, snow, sleet, fog, wind—may play a part in migration. Some of the most obvious factors include the direction from which the wind is blowing, which may help the bird to reach its destination or hinder it; extensive cloud, which obscures the stars, making finding the way more difficult; and the temperature of the destination in spring—if it is still cold and frozen, rather than warm, insect food may not be readily available.

For a species as a whole, however, climate and the changing weather patterns from year to year are the major influences that shape migratory routes and destinations. Weather requires tactical decisions by individual birds; climate involves strategic considerations that may affect the behavior of whole populations over many years.

The changing seasons

Birds are forced to migrate by the effects of the changing seasons. Areas where the habitat and weather afford a good living in the summer for the breeding birds may be untenable outside the breeding season. Many species respond to winter and summer, but others, particularly in the tropics, are governed by wet and dry seasons. The time to breed is determined by the flush of productivity that characterizes the rainy season. This burgeoning of plant and animal food may occur regularly, at a set time of the year, as in eastern Africa, or at long and irregular intervals, as in much of the interior of Australia.

Seasonality is also important for birds using traditional staging posts. In northern Portugal and northwestern Spain, many small migrants from farther north and farther east in Europe find ideal conditions for putting on weight in the fall. Insect populations are high, and there is also plenty of fruit which can be exploited.

These same areas in spring, by contrast, are cold and inhospitable and able to sustain only relatively small populations of resident birds and breeding migrants. The insect-eating migrants heading for more northerly areas shun this part of Europe in spring and return to their breeding grounds via more easterly routes.

It is interesting to speculate what would happen if these areas could support large numbers of both residents and breeding

ATROCIOUS SPRING WEATHER brings down a heavy fog (right), which although seemingly inhospitable to migrant waterfowl, does not force them to land and wait for conditions to improve. Despite the overcast sky, the birds can see that there is ample vegetation on which to feed remaining from the good fall many months before.

The survival of the species, and the continuance of its genes, means that individual birds are prepared to probe the country farther to the north to ascertain whether the next staging post is free of ice and snow and if the breeding area is clear.

For species that breed in the high Arctic, winters are long and summers short. There is therefore tremendous pressure to get back early to be able to exploit just a few extra days of possible fair weather. If winter comes early, latecomers among the breeding birds may find themselves unable to complete their nesting attempt successfully and will waste a whole year without producing any offspring. And the first pairs back have the advantage of choosing the best areas in which to breed, which are also easy to defend against those that arrive later.

migrants. Would the passage birds in the fall be able to compete with an enhanced local population for available food resources? Or would the rest of Europe suffer from a depleted population of migrants because a vital link in the migration chain had been broken?

The human element

All birds are at the mercy of inclement weather, and terrible disasters do happen to migrants on occasion. Many of these are natural, but others are the direct or indirect result of human activities. Humans have engineered enormous changes to the world's climate and environments that surpass modern concerns over global warming and damage to the ozone layer.

Huge shifts have come about through the destruction of forest and ground cover. Many tracts of the Sahara desert, for example, were productive farmland and pasture as recently as 1,600 years ago.

Perhaps the most distressing immediate losses result from the attraction of migrants to bright lights in fog, which causes death by collision. The rotating beams of lighthouses were once a hazard, but modern flashing lights are not dangerous. In many cities, the lighted windows of skyscrapers have proved fatally attractive to birds: owners of some buildings in North America employ people to gather the bodies in the morning. One scientific paper in the 1970s documented the deaths of 140,000 warblers from 135 sites.

Getting it right and getting it wrong

Although migration is usually timed to take advantage of the weather, mistakes can happen. Some Blackpoll Warblers of North America are experts at getting it right (see pp. 14 and 74–75), waiting—already laden with enough fat for their travels—until a succession of low-pressure systems passes over the Massachusetts coast at four- or five-day intervals. When, by the change in local weather, they detect one approaching, they take off and use the winds around it to help them fly out to sea where, hundreds of miles off the coast, they come under the influence of the northeast trade winds, which speed them on their journey south.

Weather and climate ②

Many birds wait for settled conditions before migrating. These often occur during extensive high-pressure systems. If the weather is overcast and wet, migration may be held up for a few days. Some strong fliers, such as shorebirds and waterfowl, are not discouraged by local difficulties and may migrate even in seemingly inclement conditions. But they often migrate in small steps (compared with the distances they are capable of traveling in one flight and the total distance they have to cover), so are not taking great risks. A bird about to set out on a long, nonstop flight, on the other hand, cannot afford to take any additional risks and is more likely to wait until the local conditions look perfect.

Migrants that misjudge conditions may die. There are, however, several steps a bird can take to increase its chances of survival if it realizes that something is going wrong. The first thing to do is to settle and wait for conditions to improve. Many birds of different species have been found in unlikely places for this reason, but for a land bird flying over the ocean, places to touch down are at a premium. This is why headlands and islands have become such prime sites to witness migration, albeit migration going wrong. It also explains the attraction of ships and oil platforms for birds in bad weather.

If a place to settle is not immediately available, birds can drift with the wind, waiting for an opportunity of salvation. They may be able to fly for many hours at a slow speed and, with strong winds to help them, cover huge distances, albeit in the wrong direction. The number of birds of many species, representing many different groups of land birds, regularly wandering across the Atlantic Ocean in both directions shows how effective this strategy can be.

These falls—sudden appearances of large numbers of migrants in the wrong place at the wrong time—are regular, if not frequent, occurrences. Scandinavian breeders are some of the birds most often affected. Autumn migrants in their millions take off from Scandinavia in apparently excellent conditions to travel southwest across the North Sea, en route for Spain and Portugal via Britain and France. But weather fronts build up in their path. Often, overcast conditions combine with a southeasterly wind that causes the birds to drift downwind too far west and onto the nearest shore. On September 3, 1965, a fall brought the Suffolk town of Lowestoft, in England, to a standstill for more than an hour as up to half a million exhausted birds of many species settled on roads and paths.

④ **WHEN THE BIRDS** do go, most travel with others and set off soon after dawn. Sea breezes may cause problems, and although the crossing is short, may take their toll.

⑤ **THE JOURNEY** through France involves a stay of three to five days at two or three roosts: earlier arrivals show the best places to feed. Torrential rain and hailstorms in August and September often cause local damage to the grapevines—and the death of any Barn Swallows unable to find shelter.

⑥ **THE PYRENEES** represent a major hurdle. Many birds skirt the edge of this huge range, but others cross it using the major passes. This may be the birds' first experience of mountains—and their fickle weather. Some are lost to thunderstorms.

Here, and around the Mediterranean, human hunters take a huge toll of small birds, but not usually of Barn Swallows.

⑦ **BIRDS CROSSING SPAIN** to reach Gibraltar have a short over-sea flight—only 8 miles (13 km) if they choose the narrowest crossing point—into Africa. Others are not so fortunate. Birds crossing the Mediterranean may have to fly as much as 300–370 miles (500–600 km), and if not properly prepared they may perish. Those taking a route too far to the west may be swept out into the wide expanses of the Atlantic and collapse exhausted into the sea or onto a ship.

❸ THE JOURNEY of at least 20 miles (35 km) across the English Channel presents the birds with their first flight over a large body of water. Many are reluctant to take this step and go a little way out to sea and then fly back again. Poor weather prevents any attempt at the crossing, and coastal roosts may hold increasingly large numbers of birds.

❷ BY THE TIME they have started to journey south, their reedbed roosts have grown in size and contain birds from many areas. In cold weather, food is at a premium, and the average weight of juveniles coming to roost may decline by more than 10 percent. Some may die of starvation; others may fall victim to predators, such as the Hobby.

❶ YOUNG BARN SWALLOWS stay in their nesting area, roosting in local reedbeds and protected from most predators by the water in which the reeds grow. Flocks often form on telegraph wires, honing their flying skills and feeding techniques, and familiarizing themselves with local landmarks and the feeding areas they will use if they survive to return next year.

❽ THE AFRICAN SHORE may be the end of the journey for the birds, since colonies of Eleonora's Falcons, spread along the coast, time their breeding to coincide with the autumn migrants moving through—ensuring a ready supply of fresh meat for the falcons and their chicks.

IT IS IMPOSSIBLE to know what will happen to migrants on their flights—but it is often possible to guess. Young Barn Swallows traveling from their nests in the British Midlands to their winter quarters in South Africa face a journey of 6,000 miles (10,000 km)—a daunting prospect for birds that weigh only $^2/_3$ ounce (20 g). This is one of the best-documented journeys of a small migrant, since there have been many banded recoveries from both Europe and Africa.

Survival of the individual is not crucial to the survival of the species. Of two parents setting out, only one will, on average, return the following year—a mortality rate of 50 percent. Of the five or six offspring that may start the journey, only one will return—a mortality rate of 80 percent.

❾ NORTHERN AFRICA holds few perils once the coast is passed. The Atlas Mountains provide varied habitats and good feeding. But this is a lull before the birds' most formidable obstacle—the Sahara desert. In places it is more than 950 miles (1,500 km) across and, except around oases, provides little opportunity for flying insect-eaters to forage.

The winds are usually in their favor but dust storms are a hazard. More Barn Swallows probably perish on this phase of their journey than on any other.

❿ IN THE SAHEL, autumn rains mean that the birds find plenty of food and can quickly regain lost weight. The journey south through the semidesert, savanna, and open woodland may be leisurely and relatively hazard free.

⓫ BIRDS THAT ATTEMPT to cross, rather than go around, the rainforest of the Zaïre basin may be in for a shock: spectacular storms daily cause the death of migrants foolish enough to overfly this rich habitat.

⓬ BARN SWALLOWS that make use of the more equable and open habitat east of the rainforest and survive face one last hazard: they may arrive in South Africa before the last of the late frosts and even snow have given way to the southern spring. In some years mortality has been high in November and early December. But, for the most part, South Africa probably reminds the birds of the best of the summer weather they were experiencing four months earlier in Britain.

How migration is studied ①

Direct observation and skillful interpretation reveal many of the secrets of bird migration.

The fact of bird migration was gradually discovered over many centuries through a variety of observations. Some 5,000 years ago, the seasonal movement of numbers of large birds over the Mediterranean island of Cyprus was taken as a signal to plant the crops. The Greek philosopher Aristotle wrote that the summer Redstarts, which are now known to leave Greece for sub-Saharan Africa, were transformed into Robins, which breed farther north and winter in Greece. Likewise, he thought that summer Garden Warblers became winter Blackcaps. Both are incorrect interpretations of accurate observations, but understandable given that the two pairs of species are similar in shape and size.

From Aristotle to Gatke

Despite these errors, Aristotle believed in migration and correctly labeled a variety of birds as migrants. It was his mythic theory of transmutation, however, that was slavishly reiterated for almost 2,000 years in the writings of other observers. Among the species he classified as migratory was the Barn Swallow, coincidentally the subject of the first successful marking experiment. The prior of a Cistercian Abbey in Germany, writing in about AD 1250, reported that a man had tied a parchment to the leg of an adult Barn Swallow with the message "Oh, Swallow, where do you live in winter?" He must have been delighted to get the message back in spring "In Asia, in the home of Petrus."

When Swedish biologist Carl Linnaeus devised an effective system for naming species over 250 years ago, he opened the way for a systematic study of bird populations and movements. Even then, there were hiccups: the letters of British naturalist Gilbert White, echoing the writings of Olaus Magnus, refer to swallows, swifts, and martins hibernating in mud at the bottom of ponds. (White may not have believed this himself, but he was not inclined to contradict his patron, who did.)

Interest in collecting bird specimens and the development of taxonomy made it easier to identify patterns in the occurrence of birds all over the world and to log major patterns of migration. In an age when the gun rather than binoculars aided bird investigation, the adage "What's hit is history, what's missed is mystery" prevailed. The skills of field identification were not well developed, and sightings were more likely to be discussed over dinner and the skin than over photographs and field notes.

Collectors came to realize that rare and interesting birds could be obtained at times of migration on remote islands and headlands and, until the 1920s, dispatched crack shots to augment their collections. One of the first places explored was Heligoland at the northeastern extreme of the North Sea. Over a period of 50 years, German ornithologist Heinrich Gatke recorded 398 species; his magnum opus, *Heligoland as an Ornithological Observatory*, was published in English in 1895.

HAULING IN THE CATCH

One of the many fanciful theories put forward to account for the seasonal changes in bird populations was that species hibernated in winter.

Olaus Magnus, Bishop of Uppsala, Sweden, although acknowledging the writings of others who claimed that Barn Swallows migrated to warmer regions for the winter, insisted that northern birds went to roost in reedbeds in the fall, allowed themselves gradually to sink into the mud, and spent the winter asleep there.

A woodcut from Magnus's *A History of the Northern Nations*, published in 1555, illustrates this notion as fishermen haul in both fish and Barn Swallows.

De Hirundinibus ab aquis extractis.

MIST NETS, INVENTED IN JAPAN, provide an efficient and safe means of catching small birds in flight. Birds caught in this way are banded and may also be weighed. They are weighed again if recaptured.

This process has demonstrated that Sanderlings (left) can nearly double their normal weight in the spring, prior to their northward migration, from $1^7/_{10}$ ounces (48 g) to a peak of $3^1/_{10}$ ounces (87.5 g). This is accounted for by a store of fat which is sufficient for a flight of about 2,500 miles (4,000 km) from South America to a re-fuelling stop on the Gulf Coast of the United States on their way to breed in the Canadian arctic.

THE SIMPLEST WAY TO TRACE A BIRD ON MIGRATION is to place on its leg a metal band (above) that includes a return address and a unique serial number. In this way, anyone finding the bird can send details of place, circumstances, date, and the serial number to the address on the band. Two dates and places are then logged.

For small birds, the chances of recovery are remote, possibly 1 in 300 or so. For larger birds such as geese and ducks, the "recovery rate" is higher: 1 in 10 banded birds are eventually reported.

How migration is studied ②

Heligoland and other headlands, as well as islands, were such prime sites for bird collectors because of their lighthouses. Many night migrants were attracted to, then disorientated by, the rotating beams and perished (modern lights, constructed differently, pose no threat). From 1880 to 1914, lighthouse keepers on the islands and headlands of northern Europe logged the birds observed and specimens killed at their lights, with counts collated in Britain.

Migration study today

Islands, headlands, and similar sites worldwide are now bird observatories. Not all are on the fringes of the continents: Long Point on the shores of Lake Erie, Canada, has long been a well-known observatory. Often they are manned by dedicated bands of local birders, but many have developed into field stations where students of migration and birdwatchers interested in the subject—and in seeing rare birds—can stay. Most observatories have a census area where the daily totals of birds of each species are recorded, and regular counts of visible migration, that is, birds that can be seen actively migrating, are logged. Coastal observatories mount regular sea watches to count sea birds passing by. Banding is also usually carried out.

Many professional studies of migration are now conducted under the auspices of universities or similar institutions and

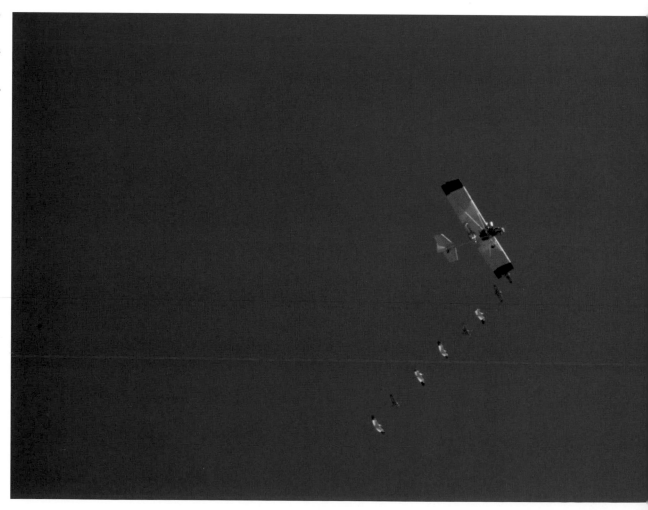

A NUMBER OF STUDIES of the behavior of migrants have been made by glider, light aircraft, and ultralight. This technique is particularly successful for studying migrants soaring over land, since it allows individual birds to be followed over considerable distances, and their height, speed, position, and the amount of flapping they do to be logged.

Snow geese (right), storks, pelicans, cranes, eagles, vultures, and several other species have been studied in this way. There are even instances of pilots going up at night to find out whether swifts were asleep while flying: they were, but only for very short periods.

One of the best studies concerned Common Cranes over southern Sweden, where individuals were continuously tracked for up to three hours at a time (see pp. 96–97).

tend to be more experimental than observational. Projects may involve altering birds' sensory perceptions—equipping racing pigeons with magnetic coils or frosted contact lenses, for example—but usually experiments are carried out on birds in captivity. In North America, birds have been placed in planetaria and exposed to different star patterns, in rooms with controlled magnetic fields, or in cages with mirrors to alter the apparent direction of the sun (see pp. 30–33).

Since the 1960s, two types of observational research have become common. The first, hi-tech, approach is to use radar to track the birds. Birds showed up on radar screens from the first (although they were initially not recognized as such and were labeled "angels"). Radar sets were later adapted for bird study, and modern antiaircraft detection equipment, suitably modified, can now track an individual bird over a range of more than 6 miles (10 km). The signal accurately pin points a bird's position, height, speed, wing-beats, and heart-beats.

The other approach, more popular in North America than in Europe, is moon-watching. This is a means of tracking the level of nocturnal migration by watching birds flying across the moon. For best results, the moon should be full and high in the sky: September and October are good months in the northern hemisphere. A pass every one or two minutes may occur at peak times (see p.15).

In the last 10–15 years, small radio transmitters have been fitted to birds, including seabirds, waterfowl, and raptors. The latest transmitters send signals to orbiting satellites and these are relayed to ground stations, allowing whole migration routes to be mapped (see p. 90–91).

FEW TECHNIQUES for studying migration have yielded such exciting results as radio tracking. When a transmitter is fitted to a bird (above), its position can be plotted through a receiver tuned in to the pulses radiated back. Transmitters are becoming smaller: many can now be glued to the tail feathers of the bird, with the antenna down the shaft. And, as battery technology improves, the range of the signal increases.

Ongoing research since the late 1980s uses satellite tracking to log a bird's position a couple of dozen times a day. Among the species tracked so far are Bewick's Swans, on their way north from the Baltic, and Wandering Albatrosses In the Southern Ocean. It has been found that the swans appear to send out scouts (see p. 26) and that the albatrosses were frequently becalmed for up to 36 hours by the weather.

Threats and conservation ①

Human activities are the greatest threat to migrants, and the only solution to their problems.

Without a doubt, the major global threat to all birds, migrant and resident, is habitat destruction. Birds have evolved over millions of years to exploit the world's rich variety of habitats. Migrant birds are particularly at risk, since they are not only likely to rely on different habitats in winter and summer, but may also use several more sites on route between them. Habitat destruction may be gross, but it can also be subtle: the replacement of a mature, natural forest by a single-species plantation can be as damaging as complete felling, altering the numbers and composition of both insect species and other forest dwellers. Most of the problems caused by habitat destruction are due—directly or indirectly—to human activities.

The importance of water

Bird populations have been devastated by wetland modification. Often the characteristics of the wetland that are most important for birds are those that are most inconvenient for people. Seasonal waters with large areas of marsh are often vital for birds. Yet in the tropics, these areas may be turned over to rice paddies; in temperate regions, they might be ponded and banked to concentrate the water into smaller tracts for fish farming, for example, with the reclaimed area turned over to agriculture.

Estuaries have been used for centuries for trade and industry. It is difficult today to think of the Hudson River in New York or the port at Rotterdam as important places for shorebirds, but they once were. Not only are the physical characteristics of such sites—the buildings, wharves, and docks—unsuitable for birds; industrial effluent may also render the estuary sterile. Ongoing estuary development means the loss of staging posts and wintering grounds. Land-reclamation schemes, by contrast, can encourage birds. The polders in the Netherlands, for example, have provided marvelous habitats for a wide variety of water birds in the sump areas that are not destined for agricultural use.

Water may be extracted for industrial, domestic, and leisure use. Marinas destroy wetland sites; hotels and golf courses eliminate primary habitats; and the sheer weight of tourist numbers can be an overwhelming disturbance.

Field and forest

When large tracts of the northern hemisphere were forested following the retreat of the glaciers at the end of the last ice

OIL ON THE SURFACE OF THE SEA has left huge numbers of birds dead (below). The 1993 wreck of the Braer, off the Shetland Islands, killed fewer birds than usual in such disasters since storms dispersed the oil. Smaller spills have accounted for much larger losses.

Spills tend to have most effect on wintering sea birds, and because they are traditional in their choice of wintering area, it may take a decade or more for populations to recover.

TROPICAL RAINFOREST is rapidly disappearing (right). Many migrants winter in secondary forest and woodland, which are also being modified or destroyed.

Secondary forest used to result regularly from slash-and-burn operations: after two or three years of cropping, the land returned naturally to forest that generalist migrants could exploit. Today, rainforest destruction is often total, with soil erosion or permanent grazing rendering vast tracts useless for birds.

age, woodland birds reigned supreme. Deforestation has had a devastating effect on birds in the last few thousand years. Some of these species may have benefited from more open habitats, but most have suffered considerable declines. Recent changes in temperate regions have included replacing native forest with plantations of exotic monocultures. This inevitably results in a restricted range of bird species and a lower biomass (see p. 73). The planting of Australian eucalyptus trees in Mediterranean

Mallard

HUNTERS IN THE U.S. fund conservation by buying Duck Stamps (see pp. 54–55). The stamps fund research into threatened populations and enable the planting of special fields for the ducks, to limit damage to crops.

areas has affected birds in summer and on migration.

Every new road, factory, office, and house is likely to displace breeding birds. Each improvement to agriculture is designed to increase yields of crops, thereby leaving less room within the crop for the weeds and pests that the birds would otherwise exploit. In Britain alone, 15 species which fed largely on seeds in the agricultural environment have declined over the last 20 years, several by more than two-thirds.

Threats and conservation ②

Pesticides cause serious problems. Direct kills of birds tend to occur in developing countries where safety rules may be lax and where chemicals no longer approved in the developed world may be dumped. There is also an indirect, and probably more serious, aspect of pesticide contamination—the global build-up of persistent chemicals, such as DDT. Chemicals have also affected the ozone layer, which in turn affects the food supply. The greenhouse effect may prove damaging: if sea levels rise, estuaries and other prime sites will be lost.

Hunting and caging

Migrants, including traditional quarry birds such as ducks and many other species, are often the targets of hunting. Small migrants like Blackcaps are at risk in the Mediterranean, and illegal trophy hunters shoot Honey Buzzards in southern Italy. In Senegal it is thought that as many as 25 million birds may be taken and caged each year—more than the total number of birds banded worldwide in a year. Caging is rife, too, in Southeast Asia but here birds are also eaten and used in traditional medicine.

Redressing the balance

National and international conservation bodies are acting on their own, as well as together and with governments on behalf of migratory birds. Their projects are often locally based, but their effects may be far-reaching. (Schemes for specific species are mentioned where appropriate.) Designation of estuaries as reserves is vital, but other wetlands are crucial to other species—there are networks of reserves in North America, Europe, and Japan to protect wild swans. Similar networks exist for White Storks and cranes, and also allow teams of scientists to study the birds.

BirdLife International is the umbrella organization which coordinates and provides a focus for much of this work. Its hundreds of projects include providing identification guides to Portuguese children to try to change the culture from one of bird hunting to birdwatching; the foundation of DHKF—the Society for the Protection of Nature—in Turkey, where there are 80 important bird areas, 50 unprotected; and the protection of the Hadejia-Nguru wetlands in Nigeria from damming and irrigation, which would destroy the wintering area of huge numbers of migrants.

BANDING BIRDS may be the most effective way to monitor bird poulations. Identifying birds as individuals allows research into migration, survival rates, and population. Only through this type of research can conservation efforts be focused. Most adult birds are caught for banding in mist nests (left). Experienced banders ensure that no bird is harmed during this procedure.

MANY HUNTERS harvest only surplus birds from well-studied populations (right); others kill millions of migrants each year with no thought for the local ecology. In some areas, birds are an important part of the local diet. More often, the slaughter is a macho occupation designed to increase status. On Malta, conservationists are winning over the children to reduce the importance of hunting in the culture. It is a long, slow task, but attitudes do appear to be changing.

North American migrants

The Americas, and North America in particular, are crisscrossed by the paths of enormous numbers of migrant birds. Long before Europeans reached the continent, the North American First Nations peoples relied on migrant waterfowl and sea birds for food at certain times of the year. Today, the movements of these birds are equally important to hunters, and ways to allow sport without detriment to bird populations have been researched and developed. Indeed, the American model of treaties to protect birds, introduced in the 1920s, has paved the way for international conservation agreements.

North America stretches from the high Arctic to the tropics and provides the breeding grounds for millions of migrants—some also found elsewhere, others unique to the New World. The continent is home to some of the world's largest flying birds, such as swans and cranes, and its smallest—members of the hummingbird family. Its native birds range from those that travel only a few miles, if at all, to long-distance migrants that may journey from the far north across the equator and beyond.

Sandhill Cranes breed in isolated pockets across northern North America and, with stopovers en route, winter in the southern states and Mexico.

Patterns of migration

The geography of North America exerts the greatest influence on the continent's migrants.

North America is made for migrants. Its broad expanses of tundra and boreal forest and myriad wetland habitats lie within latitudes where summer days are long and food abundant, but where winter nights are also long and suitable food is minimal. To exploit the exceptionally hospitable environment, birds must retreat from the inhospitable. About 80 percent of the species and 94 percent of the individuals that breed in the northern coniferous forests migrate to the tropics. In the deciduous forests of eastern North America, 62 percent of the breeding species and 75 percent of the individuals migrate. In the central grasslands, 76 percent of breeding species and 73 percent of individuals are migrants. Among more western and southern breeding populations, the percentage of migrants declines greatly.

No more than about 130 of the nearly 650 species that nest in the United States or Canada are nonmigratory, and many of these undergo local seasonal movements—from higher to lower elevations in mountains or dispersing from nesting to non-nesting areas, for example. They include many woodpeckers, grouse, and quails, and a few owls. Others such as the Phainopepla, Wrentit, and Limpkin are limited to warmer areas and have no need to migrate.

The migrations of many species—including most of the continent's waterfowl and many sparrows and finches—are completed entirely within temperate North America. But some 330 temperate species (more than half the total) winter in the New World, or Neo-, tropics. Among them are most North American swallows, flycatchers, thrushes, vireos, wood warblers, and tanagers, and many raptors and shorebirds.

The continent's geography makes a vital contribution to North America's migratory pathways. Waterfowl biologist Frederick Lincoln, recognizing this fact, in 1952 formulated a generalization referred to as the "flyway concept." His Pacific, Central, Mississippi, and Atlantic flyways became the focus of study of hunted waterfowl populations. As with many generalizations, the flyway concept is not so simple. Through studies of banded individuals, it is now known that birds nesting in one flyway may move east to west or diagonally to migrate down another. It is also clear that some species—or even some populations—do have narrow migration pathways, but many follow broad routes that do not conform to any flyway.

For some birds, the Florida peninsula is a guiding finger to launch southbound migrants along the stepping stones of the Greater and Lesser Antilles to the rich wintering areas of South America. To others, subtropical Florida, the Bahamas, and the islands of the Caribbean provide winter retreats. For northbound migrants, Florida is a beckoning finger and its Atlantic and Gulf coasts dispersing arteries.

The broad east–west expanse of the northern coast of the Gulf of Mexico is a barrier to some southbound migrants, deflecting them to the east or west for safe passage to the south. For others, it is merely a gathering place where migrants build up reserves of fat and wait for favorable winds before striking out across 500 miles (800 km) of open water for the coast of Yucatán and beyond.

✪ HOT SPOTS

1. Point Reyes Bird Observatory, California
2. Bear River NWR, Utah
3. Hawk Ridge, Minnesota
4. Whitefish Point, Michigan
5. Cedar Grove, Wisconsin
6. Point Pelee NP, Ontario
7. Derby Hill, New York
8. Plum Island/Parker River NWR, Massachusetts
9. Martha's Vineyard, Massachusetts
10. Cape May, New Jersey/Delaware Bay, Delaware/ Hawk Mountain, Pennsylvania
11. Chesapeake Bay, Maryland
12. Cheyenne Bottoms, Kansas
13. Mazatlán to San Blas, Mexico
14. Live Oak/Aransas NWR, Texas
15. Galveston Is., Texas/ Cameron, Louisiana
16. Dauphin Island, Alabama
17. Key West, Florida
18. Dominica, Lesser Antilles
19. Panama Canal Zone

Coastlines, mountains, and rivers are attractive to migrants, both for the energy savings that can be achieved by following their course and for the abundant food supplies. For these reasons, many North American hot spots are along or close to such sites.

Cheyenne Bottoms is important for shorebirds while Cape May, Whitefish Point, Hawk Mountain, and Panama are prime sites for watching migrant hawks. Dauphin Island and Cameron are good sites in spring. For a week around April 21 each year, vast numbers of trans-Gulf migrants, exhausted by their flight, replenish energy reserves at Dauphin Island.

SOME FIVE BILLION LAND BIRDS from 500 species annually leave their North American breeding grounds for winter quarters farther south. Most reach Central America; a few travel as far as southern South America.

BOBOLINKS make one of the longest migrations of any North American land bird—those breeding in Canada fly more than 5,000 miles (8,000 km) to the grasslands of southern Brazil, Uruguay, and northern Argentina.

THE SORA has one of the longest migration routes of any member of the rail family—some birds travel almost 3,000 miles (4,800 km) from their North American breeding marshes to the Caribbean islands.

NOWHERE ELSE IN THE WORLD do the avenues of migratory access and retreat appear as clearly laid out as in North America. The landscape of the continent funnels southbound migrants to areas of winter concentration and disperses northbound migrants to the broad expanse of suitable breeding habitat.

The Atlantic and Pacific coasts are attractive to migrants. The abundant food resources concentrated along the shorelines support itinerant shorebirds. The difference in temperature between land and sea results in constant breezes to buoy up migrants. And the coastlines themselves provide continuous north-south leading lines (see pp. 34–35) to aid navigation.

The continent's major mountain ranges—the Sierras, Rockies, and Appalachians—extending more or less north-south, and the updrafts generated as winds are deflected by their steep slopes both direct and facilitate the passage of migrants. And the Mississippi River and its tributaries provide yet another north-south conduit rich in food resources and potential stopover sites.

Major route

Ducks

People who hunt waterfowl in North America also raise funds for conservation.

The migrations of ducks have probably been the subject of more study over a longer period than those of any other North American birds. Their annual arrivals and departures, and the habitats they favor, have been well known to hunters along the Atlantic coast for more than 300 years. Many North American waterfowl winter in protected Atlantic, Pacific, and Gulf coastal areas; others are concentrated in the wetlands of the lower Mississippi. Substantial numbers of some species winter in Mexico.

Declining populations of Wood Ducks and other waterfowl, and recognition of the potential to learn about populations and their movements, prompted the U.S. Biological Survey to initiate a formal bird-banding program in the 1920s. Knowledge of duck movements led to the development of the "flyway concept" (see pp. 52–53) and to hunting limits and seasons based both on these flyways and on population changes along them.

At the same time, natural habitats were (and still are) being lost to wetland drainage, human developments in coastal areas, and contamination of wetlands with agricultural and industrial chemicals. The damming and channelizing of rivers and streams, and diversion of whole rivers (such as the Kern River in central California) for irrigation are also having serious effects. Loss and degradation of habitats are the major problems facing ducks and other migrant waterfowl.

The Duck Stamp (see p. 47) was introduced in an effort to compensate for these habitat losses. Over $670 million has been raised from the stamps, and more than 5.2 million acres (2.2 million ha) of waterfowl nesting, migration, and wintering habitat have been acquired for conservation. These lands include some 545 National Wildlife Refuges and 162 additional areas administered by the U.S. Fish and Wildlife Service, in all 50 states.

Nonhunters, too, are encouraged to purchase the stamps because of their value to conservation. An annual competition to provide the stamp artwork generates further interest, and several states also sell their own stamps to raise funds for conservation. These efforts are supplemented by the acquisition and management of habitats by Ducks Unlimited, a conservation organization funded largely by contributions from sportspeople.

Ecologists are still debating the long-term effects on duck populations of one aspect of human interference in the landscape: the construction of reservoirs, which has had a tremendous impact on the distribution and seasonal abundance

Blue-winged Teal

ABOUT 100 MILLION North American ducks migrate each fall, although as a result of winter losses, spring migrations are smaller. Some 62 million ducks breed each year.

Male Blue-winged Teals (below) are among the earliest southbound migrants, arriving in the southern states in late August and September. Females and young, however, leave the breeding areas later, at times pushed south as their feeding grounds ice over.

In the fall, flocks of Blue-winged Teals often contain 100 or more birds. By contrast, pairs form in the wintering areas and gather into flocks of fewer than 30 for the northward journey.

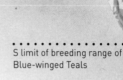

S limit of breeding range of Blue-winged Teals

BLUE-WINGED TEAL

BLUE-WINGED TEAL
Anas discors
Wingspan
24–31 in / 61–80 cm
Weight
13–15 oz / 370–425 g
Journey length
0–7,000 miles / 0–11,200 km

OLDSQUAW
Clangula hyemalis
Wingspan
26–31 in / 66–80 cm
Weight
29–33 oz / 815–930 g
Journey length
300–3,000 miles /
500–5,000 km

ESTIMATING NUMBERS and understanding the distribution of migrating sea ducks can be difficult because these birds often feed well away from land. Range maps of the Oldsquaw, for example, rarely indicate that the species winters in the northern Gulf of Mexico. Yet it is a regular winter visitor near barrier islands about 10 miles (16 km) off the coast of Mississippi. Use of open-water marine habitats makes birds vulnerable to oil spills, and there is mounting evidence that populations are in decline.

These birds are unusual in that their summer and winter plumages are strikingly different. Although they nest on the tundra of northern Alaska and Canada, nonbreeding birds, those that fail to nest, and some who leave their young before they fledge make a molt migration (see p. 25) to areas north of the breeding grounds before migrating south for the winter.

Oldsquaw

0°
Most common
in Galápagos
Oct–Mar

Route not known

of waterfowl in North America. In Kansas, for example, midwinter duck populations increased eightfold between 1949 and 1973. Reservoirs are attractive to waterfowl because they are usually both sufficiently large and deep to remain at least partially ice free well into winter. As a result, migrants winter farther north or at least take longer over the journey to more southerly wintering areas. Such shifts in wintering populations have been called "short-stopping," since the birds stop short of their traditional destination.

Most North American ducks migrate north–south, but the White-winged Scoter has an east–west migration. This species breeds in the northwestern quarter of North America, from central Canada to Alaska, yet 60 percent of these sea ducks migrate east to winter on the Atlantic and only 40 percent migrate to the Pacific coast. This pattern may reflect a past geographical separation of populations on opposite sides of North America by the mid-continent glaciers during the Pleistocene ice ages (see pp. 8–9).

Birds of prey

Half a million migrating raptors are funneled through the Isthmus of Panama each year.

Fifty years ago, hunters located areas where migrant soaring birds of prey were concentrated along ridges, such as the Kittatinny Ridge in Pennsylvania, and points of land in coastal areas, such as Cape May, New Jersey. They then lined the roadsides to shoot the birds that flew past. Today, binoculars have replaced the guns as thousands of human visitors are drawn each year to Hawk Mountain on the Kittatinny Ridge and to Cape May to study or simply enjoy the migrants in flight.

Migrant hawks, falcons, and vultures take advantage of rising air currents pushed up along ridges and of constant breezes in coastal areas. Concentrations of shorebird prey, and the leading lines created at the interface between land and water or mountain and valley, also contribute to the value of these unique, narrow migratory routes.

To appreciate fully the significance of such routes, it is important to visualize eastern North America before the Europeans arrived: dominated by forests from the Atlantic to west of the Mississippi. Since stands of trees absorb the sun's heat, there are no thermals over forests. Today, fields and roads can create thermal highways for the birds, but when the land was forested, soaring birds found energy savings only along ridges and shorelines.

The attraction of rising air over the land, coupled with the dramatic narrowing of the land bridge between North and South America at the Isthmus of Panama, concentrates thousands of migrating hawks and vultures as they are funneled to and from their wintering areas each year. Other points at which large numbers of migrating raptors can be seen include Whitefish Point in Michigan, where a narrow corridor separates Lakes Superior, Michigan, and Huron; Hawk Ridge, Minnesota, where raptors skirt the western end of Lake Superior; and coastal areas such as Rio Grande Valley in Texas.

Two further factors affect raptor migrations: diet and habitat change. Red-tailed and Broad-winged hawks must move south in cold weather, when there are no thermals to use while hunting, and when snow covers their hunting grounds, making mammal prey less common. By contrast, Sharp-shinned and Cooper's hawks, which prey on birds and frequent sheltered forest areas, may remain farther north. Prior to the 1950s, Swainson's and Broad-winged hawks rarely wintered in North America; today they are found with increasing regularity in the southern states.

DIFFERENTIAL MIGRANTS

Not all raptors migrate, and those breeding nearer the equator are less likely to be migrants than those farther north or south. This holds true even within a species. The American Kestrel (left) is found throughout most of North and South America and in the West Indies. Northernmost populations—breeding into Canada and Alaska —are migratory, traveling as far as Central America. Populations in the southern United States are often resident, augmented in winter by migrants from the north. The unique Cuban race is also resident. Migrant visitors to Cuba tend to winter along the flyways, a habitat not favored by the residents. There are also gender differences in this species' movements. Some males winter in the colder parts of the range, subsisting on House Sparrows and mice captured near human habitations. Migrant males return to the breeding grounds earlier than females to establish themselves in the best territories.

AMERICAN KESTREL
Falco sparverius
Wingspan
21–23 in / 53–58 cm
Weight
4–4¼ oz / 110–120 g
Journey length
0–3,700 miles /
0–6,000 km

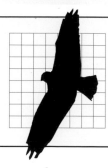

SWAINSON'S HAWK
Buteo swainsonii
Wingspan
4 ft 2 in / 1.28 m
Weight
2 lb–2 lb 6 oz / 910–1,070 g
Journey length
3,750–7,500 miles /
6,000–12,000 km

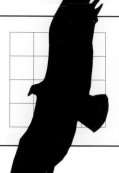

TURKEY VULTURE
Cathartes aura
Wingspan
5 ft 7 in / 1.7 m
Weight
3 lb 5 oz / 1.5 kg
Journey length
0–3,750 miles /
0–6,000 km

THE TURKEY VULTURE'S relatively heavy wings, and its habit of nesting later than its southern cousin, the Black Vulture, may be adaptations that allow it to migrate farther. Thousands of Turkey Vultures are funneled through Central America, at times in the company of Swainson's and Broad-winged hawks.

Turkey Vultures are strong fliers, passing through Middle America on the northbound journey in late February to early March and reaching the breeding grounds in the northern U.S. in mid to late March. So punctual are they that the citizens of Hinckley, Ohio, hold a festival to celebrate the return of the "buzzards" each March 15.

80° N

60° N

N limit of wintering range of Turkey Vultures

40° N

S limit of breeding range of Swainson's Hawks

⊛ Rio Grande
● Valley NWR

A few Swainson's Hawks winter in S Florida

⊛ Veracruz

20° N

Balboa ⊛

A few Swainson's Hawks winter in southern U.S. and Central America

0° N

Turkey Vulture

N Argentina is major wintering area of Swainson's Hawks

20° S

SWAINSON'S HAWKS typically migrate in huge flocks and travel farther than any other North American hawk. These birds of the plains spend their summers in the North American grasslands and winters in the pampas of Argentina.

This hawk was once intimately associated with the swarms of locusts on the pampas. A decline in locust populations may be responsible for its recent altered migration patterns and reduced wintering concentrations here.

40° S

Turkey Vultures are resident in S America; a few migrants in extreme S

60° S

Swainson's Hawk

Cranes

Habitat loss and shooting take their toll, but migrant cranes suffer most from power lines.

Only two of the world's fifteen species of cranes occur in the New World: the Sandhill and Whooping cranes. Although both are limited in their migrations to North and Central America, some of those migrations are truly spectacular in terms of distance traveled, precision of return to the same areas, and magnitude of congregations at some stopover and wintering areas. In active flight during migration, cranes may "cruise" at up to 45 mph (70 km/h); they also regularly take advantage of thermals and air currents by gliding (see pp. 16–17) over considerable distances.

Both North American crane species are birds of open, grassy, wetland areas that frequent prairies for feeding. They require open ground partly because of their size but also due to their characteristic behavior of running a few steps with outstretched wings as they take flight. As with many wetland birds, North American cranes have suffered from habitat losses, and like many large birds they have been the victims of hunting and indiscriminate shooting because they are big, easy targets. Close monitoring of migrant Whooping Cranes, bans on hunting in areas where they stop to rest, and the birds' adherence to traditional migration routes have all helped to reduce losses. Yet they still occur. The high-tension power lines that cross the prairie migration routes pose one of the most serious threats to migrant North American cranes. Low-flying cranes collide with the lines and, if they are not killed, usually suffer broken wings, becoming easy pickings for predators.

Whooping Crane

Wood Buffalo NP

S limit of breeding range,
N races of Sandhill Cranes

Monte Vista NWR

Aransas

WHOOPING CRANES ride the thermals to migrate by day, but at times may begin their travels before dawn or continue them after dark.

These birds are opportunists, feeding on a variety of plant and animal food taken from open land and shallow waters.

FROM CANADA TO TEXAS

The known population of Whooping Cranes had declined from 1,500 birds in 1850 to 15 by 1941. By 1946, it had risen to 25, only to slip back to 21 in 1952. By 1958, the population wintering at Aransas National Wildlife Refuge on the Texas coast had grown to 32; by 1965, 44. With conservation efforts, the population reached 138 in 1988 and 220 in 2005.

In 1954, after years of effort, the nesting area of these birds was found, by chance, in Wood Buffalo National Park, Canada. Thus their annual migration to Aransas is about 2,500 miles (4,000 km), flown in segments of 185–300 miles (300–500 km), with several days en route at staging areas in Saskatchewan and Nebraska.

Since 1975, Whooping Crane eggs have been placed in Sandhill nests at Grays Lake National Wildlife Refuge, Idaho, in the hope that any young Whooping Cranes would follow the Sandhills on migration and ultimately breed at Grays Lake. Both species do now migrate together to Bosque del Apache National Wildlife Refuge, New Mexico, but Whooping Cranes do not yet breed at Grays Lake.

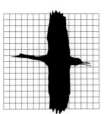

WHOOPING CRANE
Grus americana
Wingspan
6 ft 7 in–7 ft 7 in / 2–2.3 m
Weight
14 lb 2 oz–16 lb 1 oz /
6.4–7.3 kg
Journey length
2,500 miles / 4,000 km

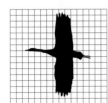

SANDHILL CRANE
Grus canadensis
Wingspan
5 ft 11 in–6 ft 11 in / 1.8–2.1 m
Weight
6 lb 6 oz–12 lb 13 oz /
2.9–5.8 kg
Journey length
0–2,500 miles / 0–4,000 km

Sandhill Crane

ALTHOUGH THEIR SCATTERED POPULATIONS at prairie potholes and marshes across Canada and the northern U.S. seem small and isolated, on migration Sandhill Cranes flock together at staging areas in incredible numbers.

Along the Platte River in Nebraska, upward of half a million of these birds may congregate on migration each spring before dispersing to their breeding grounds. The din produced by this number of birds, with their grating, rattling calls, can be deafening.

atte River/Cheyenne Bottoms

Platte River/
Cheyenne Bottoms

Known breeding and wintering areas of Whooping Cranes, with migration: routes, pre 1922

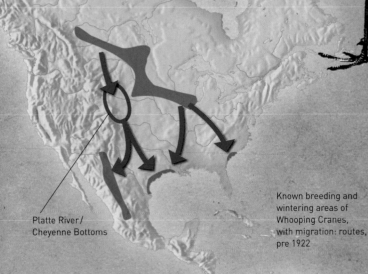

It is likely that crane populations have always been somewhat fragmented as a result of the natural distribution of suitable habitats. Today they are much more so. Sandhill Cranes migrating through the prairie provinces and grain belt of the Midwest once fed on the seeds of prairie grasses and on grasshoppers. Today they fatten on wheat, corn, and barley. Often the cranes' food is waste grain from already harvested fields, but they do also take food from, and cause damage to, unharvested fields near their roosts.

By far the largest wintering concentrations of Sandhill Cranes are found in the high plains of western Texas and eastern New Mexico. These are Lesser Sandhill Cranes, the smallest of the races and those that nest farthest north. Winter flocks of more than 100,000 individuals are not uncommon. In addition to those visiting the southern high plains, some Sandhill Cranes migrate to similar habitats in central Mexico. Many Lesser Sandhill Cranes come from the Hudson Bay region of Canada, others from as far west and north as western Alaska or eastern Siberia.

The larger Greater Sandhill Cranes nest in the northern United States and southern British Columbia. Greater Sandhill Cranes from western areas join Lesser Sandhill Cranes to winter on the western high plains. Those from the Great Lakes states migrate through the center of the continent east of the Mississippi to the Gulf States, particularly Georgia and Florida, where they are sometimes seen with the resident cranes.

In contrast to northern breeders, these southern cranes winter in their breeding areas. They include the endangered Mississippi Sandhill Crane, which now numbers about 100 birds sandwiched into a pocket of suitable habitat in Jackson County in otherwise urban coastal Mississippi. The Florida Sandhill Crane seems to have fared better on the wet prairies in the center of the state and also feeds regularly on lawns and golf courses. Little is known today of the status of the resident Sandhill Cranes of western Cuba and the Isle of Pines.

Plovers, gulls, and terns

Some archetypal sea birds migrate over a sea of grass, not water.

Contrary to popular belief, there is no such creature as a "seagull." Although many gulls and terns are birds of inshore waters and some rarely leave the open ocean, others, such as Franklin's Gull, nest far from any ocean and during migration cross the "sea" of prairie grasses and farmland that characterizes the North American Great Plains—the antithesis of "seagull" habitat.

The power of flight and the ability to migrate give a species freedom to exploit the best habitats for its needs at the optimum times. Given the mid-continent nesting area of Franklin's Gulls, it is reasonable to suppose that they would migrate south to the Gulf of Mexico. In fact, they cross Mexico to winter along the Pacific coast of Central and South America since that is where the best conditions are to be found. The cold waters of the Pacific offer a greater source of food than the warmer waters of the Caribbean and western Atlantic.

Bonaparte's Gulls, too, nest in a seemingly surprising habitat: northern coniferous forests. Like Franklin's Gulls, these birds migrate south through the middle of the continent, but they winter in more "traditional" gull habitats along the Mississippi and other rivers, where they stay until forced south by the rivers' freezing.

Herring Gulls, which nest along the northern Atlantic coast, find acceptable winter quarters closer to home, but illustrate a pattern common to many other birds: young and adults have different migration strategies. Young birds often migrate to the southern states or the West Indies but, as they mature, they winter closer and closer to their nesting areas. For this reason, birders along the Gulf coast may only get to know the Herring Gull in its immature plumage.

American Golden Plovers, nesting in the northern tundra and wintering in the grasslands of Argentina, typify yet another migratory strategy. Adults leave the nesting area before the juveniles and migrate southeast to the Atlantic coast, then south over the ocean to South America. The later-departing young migrate south through the Mississippi valley, the route taken by all birds on the spring flight north.

Like gulls, terns are opportunists. Not much larger than a swallow, the Interior Least Tern follows major rivers on migration in North America and takes advantage of sandbars for nesting. Other populations use similar habitats along the shores of the Pacific, Atlantic, and Gulf of Mexico.

In part as a result of channelization and damming of rivers, Interior Least Terns have declined to the extent that they are now considered endangered. Habitat destruction has similarly limited California Least Terns. The extent to which these populations are separate, however, is not known. Banding has shown that there is some exchange of Gulf coast and inland birds. More detailed studies of migration routes, wintering areas, and movements between populations are urgently needed.

MIGRATORY VAGRANTS

Killdeers in the north of the range are migrants; at middle latitudes they may remain through mild winters. In the Gulf States, breeding birds are often resident, some maintaining pair bonds and territories year-round.

These birds often stray from their usual migration routes and have been seen in western Europe, Greenland, and Hawaii. Such vagrancy is no doubt a result of their normal long migrations and association with coastal habitats on migration and in winter. Killdeer nests, by contrast, are usually not near water.

Franklin's Gull

ITS CHOSEN NESTING AREA, together with its pale gray and white plumage, have earned Franklin's Gull (right) the popular name of "Prairie Dove."

These birds nest beside marshes and small lakes and feed on surrounding farmland, eating agricultural pests such as wireworms and cutworms. Although their breeding range is restricted, they are numerous—nesting colonies may contain thousands of pairs.

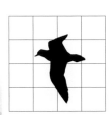

KILLDEER
Charadrius vociferus
Wingspan
18–22 in / 46–56 cm
Weight
3–3½ oz / 90–100 g
Journey length
0–6,000 miles / 0–10,000 km

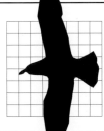

FRANKLIN'S GULL
Larus pipixcan
Wingspan
35–37 in / 89–94 cm
Weight
8–12 oz / 230–340 g
Journey length
1,900–5,000 miles /
3,000–8,000 km

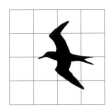

ROYAL TERN
Sterna maxima
Wingspan
43 in–45 in / 1.08–1.14 m
Weight
1 lb / 450 g
Journey length
0–5,000 miles / 0–8,000 km

Route not known

60° N

40° N

Salton Sea, California

Royal Terns breed as far N
as Maryland

20° N

Franklin's Gulls
occasionally winter
in Hawaii

Royal Terns breed S Cuba,
Jamaica, Puerto Rico,
Aruba, Curacao, Trinidad

Both birds are regular
migrants in Galápagos

0°

20° S

Non-breeding

40° S

Franklin's Gull

Breeding pop. here

Distribution of Royal Terns

Major migration route N
and S is through Greater
and Lesser Antilles

Majority of African
breeders migrate S along
coast and winter from
Senegal to Angola

THE APPEARANCE of Royal
Terns in western Africa—
and, rarely, in Britain—
prompted the belief that
some North American
breeders must cross the
Atlantic on migration.
Then, in the early 1960s,
nesting colonies were
found in Mauritania,
Senegal, and The Gambia.
It is still not known whether
the two populations are
entirely separate.

Winter sightings in
Florida of adults feeding
fledged young document
that adults and their young
migrate together.

NOV DEC JAN
OCT   FEB
SEP   BREEDING   MAR
AUG   APR
    MIGRATION
  JUL JUN MAY

Royal Tern

Nightjars

Hibernation is a viable alternative to migration for one bird.

The nightjars of North America are varied in their migratory and wintering strategies. Chuck-will's-widows and Whip-poor-wills of eastern North America are often solitary nocturnal migrants, but may travel in loose flocks. Common Nighthawks migrate by day or night in loose flocks. The Common Poorwill comes as close as any bird to hibernation, with northern populations migrating and southern populations going into torpor during periods of inclement weather. The Common Pauraque, limited to the warmer areas of northern South America and north into Central America and Texas, has no need to migrate.

As aerial insect-eaters, all migrant nightjars are capable of feeding en route. Chuck-will's-widows also consume warblers, small swallows, hummingbirds, and sparrows, presumably captured during nocturnal migration. This ability to eat small birds, along with their larger size, probably allows a few Chuck-will's-widows to spend the winter in the southern United States.

The spring migrations of Chuck-will's-widows and Whip-poor-wills in the eastern United States are punctuated by their characteristic calls during stops of one to a few days as they progress northward. The impression of "migratory flocks" results from the sudden appearance of many calling birds in an area; it may be that the birds' flights are simply timed to occur simultaneously.

A second factor that contributes to the impression of flocking is the response of birds when they encounter a barrier, such as one of the Great Lakes. They follow the land as far as they can and, if they reach the tip of a peninsula—Point Pelee, Ontario, for example—they may pause for a few hours to a few days to wait for favorable winds before attempting to cross the lake. This allows many individuals to congregate, although whether the "flocks" departing from such areas are cohesive is not known.

THE COMMON NIGHTHAWK is known colloquially as the "bull-bat" because of an imagined batlike quality to its flight and the "bull-like" roar made by its wings when it pulls out of a dive.

The migrations of these birds take them to and through the West Indies, en route to their wintering grounds in Central and South America. They regularly migrate in large numbers, often in flocks of thousands. Their diurnal hawking for insects, migration in flocks, and the common name "hawk" worked against these birds in the late 19th and early 20th centuries, when they were shot for sport and food, and as predators.

Three factors have contributed to their recovery: a ban on shooting; recognition of their value in the control of populations of insects classed as pests; and the species' ability to make use of gravel rooftops as nest sites.

Whip-poor-wills are not present in U.S. and N Mexico in winter

Whip-poor-wills occasionally winter on the Gulf coast

Whip-poor-wills are present in S Mexico and Central America in winter only

Great Plains race *sennetti*

Common Nighthawk

♀

Eastern N American race *minor*

♂

COMMON NIGHTHAWK
Chordeiles minor
Wingspan
21–24 in / 54–61 cm
Weight
2⅓ oz / 62 g
Journey length
2,500–6,800 miles /
4,000–11,000 km

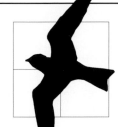

WHIP-POOR-WILL
Caprimulgus vociferus
Wingspan
18–19½ in / 46–50 cm
Weight
1¾–2 oz / 51–55 g
Journey length
300–3,750 miles /
500–6,000 km

COMMON POORWILL
Phalaenoptilus nuttallii
Wingspan
14–16 in / 36–40 cm
Weight
1¾ oz / 52 g
Journey length
0–2,500 miles / 0–4,000 km

40° N

20° N

0°

20° S

40° S

S limit of breeding range of
Common Nighthawks

THE SLEEPING ONE

Northern Common Poorwills migrate to southern California, New Mexico, Arizona, and into Mexico to avoid cold weather and food shortages. Poorwills from the southwestern United States, by contrast, overwinter by hibernating, hence the Hopi Indians' name for the bird, Hölchko, meaning "the sleeping one." A hibernating Poorwill with a ⅓ ounce (10 g) fat reserve can sustain itself for 100 days, whereas the same reserve sustains a non-torpid Poorwill for only 10 days. The hibernating Poorwill is capable of spontaneous arousal at temperatures near freezing but requires seven or more hours to wake up fully. This inability to respond quickly to danger is one of the major disadvantages of torpidity; their camouflage coloration and choice of sheltered recesses in which to hibernate help Poorwills survive (below).

Migratory Poorwills, too, may spend spells of cold weather in a torpid state.

UNTIL EARLY IN THE 19 TH CENTURY, the Whip-poor-will (below) and Common Nighthawk were thought to be one and the same bird. The characteristic call of the Whip-poor-will was explained as simply the nocturnal call of the Common Nighthawk.

"Father of American ornithology," Alexander Wilson, differentiated the two species and first described the gradual northward migration of the Whip-poor-will from its wintering areas in Central America to the breeding grounds of the eastern U.S. and southern Canada.

Whip-poor-will

♂

♀

NOV DEC JAN
OCT MIGRATION FEB
BREEDING
SEP MAR
AUG APR
JUL JUN MAY

Swifts, swallows, and martins

Migrant martins owe their survival to the prompt action of the citizens of New Orleans.

As aerial insect-eaters, swallows and martins, together with swifts, are able to feed on the wing; both groups therefore migrate by day and roost at night. Small flocks of swallows characteristically travel together; swifts are more likely to congregate in flocks of thousands of birds. At times, mixed flocks of migrating swallows and swifts are clearly visible. Most swallows nest, winter, and stop over on migration near water, which provides a steady source of insects and a buffer against extremes of temperature. Swifts also feed near water, but often roost in hollow trees or, since the arrival of Europeans in the New World, in chimneys well away from water.

Most North American swifts are long-distance migrants, with only the White-throated Swift wintering as far north as central California, southern Arizona, and New Mexico. The winter home of Chimney Swifts was a mystery well into this century. Each fall, Chimney Swifts gathered in flocks of several thousand and then disappeared. In an effort to learn more of their movements, more than 375,000 were banded at several stations in the late 1930s and early 1940s. The effort paid off in 1944 when 13 bands from Chimney Swifts were handed into the American embassy in Lima, Peru, as having come from some "swallows killed by Indians on the Yanayaco River," a tributary of the Amazon in northeastern Peru. Five of those 13 birds had been banded by Ben B. Coffey and Boy Scouts working with him in Memphis, Tennessee.

As the northernmost nesting swallows in North America, Tree Swallows are well adapted to coping with the sudden arrival of

S limit of breeding range of Tree Swallows, currently expanding southward

Majority this way

0°
Known wintering range

20° S

PURPLE MARTINS ARE long-distance migrants. In the 1980s large winter roosts were discovered in Brazil, and through a cooperative effort, the birds were sprayed at night with long-lasting fluorescent dye.

Banders in North America captured martins as they reached nesting areas and examined the birds under ultraviolet light to search for dye-marked individuals. Several were found, proving that birds from many areas of the eastern U.S. used the same roosts in winter.

Purple Martin

WITH ITS EXCEPTIONALLY LONG WINGS and tiny legs, the Chimney Swift cannot take off in still air from horizontal ground, since it hits the surface with its wings and cannot achieve lift. It roosts clinging to a vertical surface and, to take flight, drops from its perch. Once airborne, its flight is agile and strong.

Swifts are highly gregarious and migrate in large flocks to wintering areas in South America. In recent years, a few individuals have begun to winter in southern California.

Chimney Swift

CHIMNEY SWIFT
Chaetura pelagica
Wingspan
12–12½ in / 31–32 cm
Weight
⅞ oz / 24 g
Journey length
1,900–6,000 miles /
3,000–10,000 km

TREE SWALLOW
Tachycineta bicolor
Wingspan
12–13½ in / 31–34 cm
Weight
⅔ oz / 20 g
Journey length
600–3,400 miles /
1,000–5,500 km

BARN SWALLOW
Hirundo rustica
Wingspan
12½–13½ in / 32–34 cm
Weight
⅔ oz / 19 g
Journey length
1,550–6,800 miles /
2,500–11,000 km

60° N

40° N

20° N

EXPANSIONIST TENDENCIES

The Barn Swallow is one of the world's most widespread bird species, nesting across Eurasia and North America, where it has expanded its range in recent decades. Unlike many species adversely affected by deforestation, Barn Swallows have proven capable of taking advantage of the clearing of southern forests and the construction of interstate highways, and find good nest sites under bridges and in culverts.

This increased availability of suitable nest sites has resulted in the expansion of the birds' North American breeding range southward as far

Barn Swallow

as the Gulf coast and northern Florida. A few Barn Swallows are now wintering in southern Florida, but most must migrate farther south to the West Indies, Central and—in particular— South America.

Evidence suggests that a few migrants from North America are staying in the southern hemisphere year-round: there appears to be a nesting colony in Buenos Aires province, Argentina.

A few Tree Swallows winter as far S as Trinidad, Colombia, Venezuela, and Guyana

NESTING AS FAR NORTH as the tree line in Alaska and Canada, and wintering as far south as northern South America, Tree Swallows are long-distance migrants among swallows. Many, however, regularly winter closer to home, only moving as far as the Gulf States, Bahamas, and Greater Antilles.

These birds typically migrate by day in small flocks, feeding en route and often stopping to roost in wetland areas where they may spend the night clinging to the stems of marsh grasses.

Tree Swallow

MIGRATION
BREEDING
JAN DEC NOV OCT SEP AUG JUL JUN MAY APR MAR FEB

cold weather. By nesting and roosting in old woodpecker holes and housing provided by humans, they are insulated against cold nights. And, unlike most swallows, they will eat fruit when insects are scarce. Tree Swallows also regularly winter farther north than other swallows, many remaining in the southern United States. A few have survived winters as far north as coastal Massachusetts.

Tree Swallows generally migrate south by the most direct route, following the leading lines (see pp. 32–33) offered by major rivers, coastlines, and mountain ranges. East Coast and Great Lakes birds, for example, usually migrate along the East Coast and through the West Indies as far as Central America. Those from the Canadian prairie provinces and the north-central United States migrate through the Mississippi valley to the northern Gulf coast, with some continuing across the Gulf of Mexico to Central America. Those nesting in the Rockies often winter in Mexico. Populations in coastal California may be resident.

The spring arrivals of Purple Martins have been well chronicled. "Scouts"—typically adult males—arrive first from wintering areas in Amazonia, often appearing in nesting areas in February when freezing weather and snow can still be anticipated. Most females and second-year birds arrive a few weeks later. Early arrival gives the males an advantage in obtaining the best nest site and best mate, but mortality can be high if the weather takes a turn for the worse.

Late-summer departure from the breeding areas is less well documented. Those unsuccessful at nesting gather first, roosting near rivers and lakes where food is plentiful. They are later joined by early fledglings and their parents, and finally by those that have re-nested after failure or have attempted or completed a second nesting. Gradually, these flocks move south through river basins, coalescing into flocks of thousands. Near the Gulf coast, enormous flocks often roost at night on bridge supports where, at dusk, returning birds may be victims of rush-hour traffic. At one such roost near New Orleans, citizens successfully lobbied to have fences erected to prevent the birds from flying into the path of oncoming cars.

Hummingbirds

Nectar-sipping for food forces many of these colorful birds south in winter.

The hummingbirds are an extremely diverse and generally highly colored New World group of at least 341 species. They will eat small insects, but feed primarily on nectar, which they take by hovering seemingly motionless at flowers and sipping with their somewhat tubular tongues through needlelike, often curved bills.

Many species are tropical and, since flowers abound all year round at these latitudes, have no need to migrate. They include the Ruby-topaz Hummingbird and the smallest bird in the world, the 2-inch (6-cm) long Bee Hummingbird, which is restricted to Cuba and the Isle of Pines. Some 21 species, however, breed in North America, and, because of their diet, most must

migrate in winter. A few avoid it by going into a state of torpor to survive short periods of exceptionally cold or wet weather, in the manner of Common Poorwills (see pp. 62–63).

Male hummingbirds tend to be highly territorial and often migrate north in spring a week or more in advance of females to establish their territories for the season. Males also leave incubation of the eggs and rearing of the young to the females and begin moving south on the fall migration at least two weeks before the females and young. The flight south must take place well in advance of cold weather, not because these tiny birds cannot tolerate cold, but because of rapidly diminishing food supplies

RUBY-THROATED HUMMINGBIRDS are the only species to breed in eastern North America and also one of the most widespread hummingbirds, nesting as far north as New Brunswick and Nova Scotia and wintering from Mexico to Panama.

Males (which do in fact have brilliant red throats; those of females are usually white) leave the breeding areas first, often as early as July. Females and young, by contrast, may linger within the breeding range into October. These birds take a variety of routes to the Central American wintering areas. These include traveling through Florida and Cuba, through Mexico, or, incredibly, across the Gulf of Mexico from the northern Gulf coast to the Yucatán peninsula and beyond. For this nonstop flight of some 500 miles (800 km) the Ruby-throats put on substantial quantities of fat and may be aided by tail winds.

60° N

Spring   Fall

40° N

....
Resident and wintering pops. of Calliope Hummingbird overlap from here southwards

20° N

0°

DEC
NOV   JAN
OCT   MIGRATION   FEB
        BREEDING
SEP   MAR
AUG   APR
    JUL   MAY
    JUN

♀

Ruby-throated Hummingbird

♂

RUBY-THROATED HUMMINGBIRD
Archilochus colubris
Wingspan
4–4¾in / 10–12 cm
Weight
¹⁄₁₀–¹⁄₈ oz / 3–3.5 g
Journey length
0–3,500 miles / 0–6,000 km

CALLIOPE HUMMINGBIRD
Stellula calliope
Wingspan
4½in / 11 cm
Weight
¹⁄₁₁–¹⁄₁₀ oz / 2.5–3 g
Journey length
1,250–3,000 miles / 2,000–5,000 km

at this time of year. The return migrations of hummingbirds in spring seem perfectly timed with the blooming of hummingbird-pollinated food plants. In the southeastern United States, for example, Ruby-throated males arrive just as cross vine and red buckeye are blooming in late March.

Although many North American hummingbirds migrate to Mexico or Central America, some, such as Anna's and Costa's hummingbirds—both of which are restricted to the West Coast—

Calliope Hummingbird

TINY EVEN FOR A HUMMINGBIRD, at some 3 inches (7.5 cm) from beak to tail, the Calliope is North America's smallest hummingbird and the smallest bird of any species breeding north of the Mexican border.

It nests at a broad range of elevations from south-central British Columbia and southwestern Alberta, south through the Pacific northwest to northern Baja California, and east to western Colorado; all populations winter in central Mexico.

On migration, these birds are most common in mountain meadows, but occur in a variety of habitats, including (often in spring though rarely in fall) the streamside vegetation of the tall mesas of western Texas.

Spring migration tends to be slow as these birds keep pace with the blooming of nectar-producing flowers.

remain in the breeding areas of southern California year-round. Other species show seasonal movements other than spring and fall migration. The Broad-tailed Hummingbirds of the southern Rocky Mountains, for example, arrive in spring and breed at low elevations. Once the snow has melted at higher elevations, the birds often move up the mountain slopes to areas where there is a fresh abundance of flowers, build a new nest, and attempt to rear a second brood.

Nonbreeding Anna's Hummingbirds often wander north into British Columbia and, more rarely, to southern Alaska. Following nesting, other species too may stray far from their breeding grounds, taking advantage of food supplies wherever they can be found. In October 1984, a Green Violetear, a species that rarely occurs north of Mexico, made its way as far as an Arkansas backyard.

An amazing number and diversity of western hummingbird species now regularly winter along the Gulf coast and occasionally farther inland in the southeastern states. They seem to be most prominent in New Orleans, where yards are often planted to include varieties that bloom year-round. In some cases Rufous Hummingbirds have appeared at bird feeders—for these tiny birds, a bottle or dropper containing sugar water is the best food that humans can provide—in late summer before the summer-resident Ruby-throated Hummingbirds have departed.

Once these "foreigners" have been sighted, observers maintain their hummingbird-friendly yards and, as a consequence, keep their exotic visitors through the winter. Which route these errant birds take from their breeding areas is not known, but at least some that have been banded have returned to the southeastern states in subsequent winters, although often to quite distant sites. Among the most common of these birds are Rufous Hummingbirds, which breed in forests and on the edges of woodlands and thickets in the northwestern United States through British Columbia and as far north as southern Alaska. It may be that their adaptation to Alaska's cool and variable climate has made them particularly suited to the mild winter weather of the southeastern states.

Woodpeckers

The search for sweet sap forces sapsuckers south in winter.

Most of the world's woodpeckers are more or less sedentary. In North America, however, there are some strongly migratory species, a few that seasonally leave northern or high-elevation areas of their nesting range, and others that are somewhat irruptive (see pp. 162–65). Woodpeckers' seasonal movements can usually be related directly to food supplies.

The Yellow-bellied and Red-naped sapsuckers and northern populations of the Northern Flicker and Red-headed Woodpecker are long-distance migrants. In late summer, Northern Flickers gather in loose flocks and leave their breeding areas, migrating by day to the southern states. These migrations are no doubt a consequence of their ground-feeding habits and their inability to find food in snow-covered areas. Southern populations are resident.

Red-naped Sapsuckers breed from southwestern Canada to southern New Mexico in the Rocky Mountains and eastern Cascades. In late summer these sapsuckers leave northern and higher-elevation breeding areas to winter in the same latitudes as southern breeders, in Baja California and central Mexico.

Red-headed Woodpeckers also regularly leave northern parts of their breeding range, again largely due to food shortages. In summer these birds feed extensively on fruit and insects, but they undergo a significant seasonal shift in diet, eating more acorns in winter. This change may mean that movements of southern breeders are simply irruptive.

Short-distance migrants include the mountain-breeding Lewis' Woodpecker, which moves in loose groups to lower elevations in winter, and Downy and Hairy woodpeckers. The status of the movements of these two woodpeckers is uncertain: they may be true migration, irruption, or simply a dispersal of young birds from the breeding areas. Banding records reveal that some individuals move as much as 800 miles (1,300 km) from their site of banding in the fall. Return movements, however, are less well documented.

Yellow-bellied Sapsucker

THE YELLOW-BELLIED SAPSUCKER breeds mainly in Canada and the northern U.S., with isolated mountain populations as far south as northern Georgia. These birds depend to a great extent on the flow of sugar-rich sap, which comes to a halt in their breeding range in winter. As a result, they must move to warmer latitudes, although gender and age groups behave differently.

Male Yellow-bellied Sapsuckers tend to stay closer to home, often well within the snow belt, while females commonly migrate to the Gulf States, Bahamas, West Indies as far south as Dominica, or Central America as far south as Panama. Juveniles may move farther south than adult males.

Yellow-bellied and Red-naped sapsuckers were considered races of the same species until 1985. As a result, Red-napes in particular remain a species about which there is much to learn.

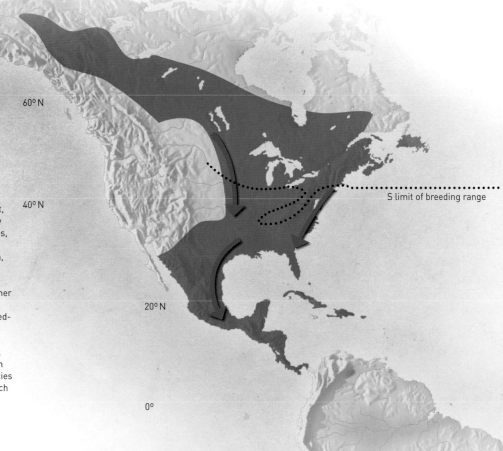

60° N

40° N

20° N

0°

S limit of breeding range

**YELLOW-BELLIED
SAPSUCKER**

Sphyrapicus varius

Wingspan
14–16 in / 35–40 cm

Weight
1¾ oz / 50 g

Journey length
0–2,200 miles / 0–3,500 km

ALTHOUGH SAPSUCKERS ARE SOLITARY, nocturnal migrants, and are rarely heard on migration and in their wintering areas, their passage, stopover, and wintering areas are well marked by characteristic sap wells. These are horizontal or vertical rows of tiny holes that sapsuckers excavate in the living tissues below the surface layers of bark. A horizontal row of sap wells indicates that a sapsucker has been there looking for sap. Once a sweet vein is found, the sapsucker begins excavating sap wells in a vertical row the better to exploit the flow. Injured or diseased trees release more sugars into their sap to combat infection, thus are more apt to attract the continued attention of sapsuckers, whose extensive sap wells can give the tree a wafflelike appearance.

Tyrant flycatchers

Wanderers return to the breeding grounds already paired.

Many tyrant flycatchers—a strictly New World group of perching birds—are unremarkable in plumage color and some are undistinguished in voice. They are often, however, champions at long-distance migration. Of the 390 or so species, 31 breed in North America and most are migratory, with some, including the Eastern Kingbird, breeding as far north as Canada and migrating as far as South America.

The Eastern and Western kingbirds and Scissor-tailed Flycatcher are open-country birds that typically migrate by day, often in flocks of 100 or more individuals. By contrast, the tiny Willow, Least, and Yellow-bellied flycatchers, which characteristically forage from perches within the cover of vegetation, migrate at night.

Eastern Kingbirds are common breeding birds in open areas of eastern North America. In southern Canada their nesting populations span the continent, from British Columbia to Newfoundland. In spite of this broad breeding range, on migration all populations funnel into the northern Gulf coast. From the Gulf States, the migration route continues to narrow, as most birds cross the Gulf of Mexico to Yucatán and low-lying areas of Central America en route to the wintering grounds of northern South America. Some may move south along the coast of eastern Mexico, a few stray to the east and appear in eastern Cuba and other islands of the Greater Antilles, and occasionally even in the Bahamas.

Western Kingbirds breed across most of North America west of the Mississippi, from southern Canada in the north to northern Mexico in the south. They winter in Central America from Mexico to Nicaragua and occasionally in Costa Rica. They also wander along the Atlantic coast from South Carolina to Florida and, more commonly, in southern Florida including the Keys.

Eastern Phoebes move much shorter distances than most other flycatchers, rarely reaching the Bahamas or Cuba. Most winter in the southern United States, and in mild winters many remain farther north, ready to move back into their breeding areas in spring. Three factors probably contribute to this tolerance of colder climates: the temperature-buffering effect of the wetland habitats that they frequent; their opportunist nesting tendencies, which have enabled them to choose sheltered nest and roost sites close to human habitations, in farm buildings, bridge supports, and porches; and their ability to consume some fruit when insect food is scarce.

VERMILION FLYCATCHER
Pyrocephalus rubinus
Wingspan
10–11 in / 25–27 cm
Weight
½ oz / 14 g
Journey length
0–2,500 miles /
0–4,000 km

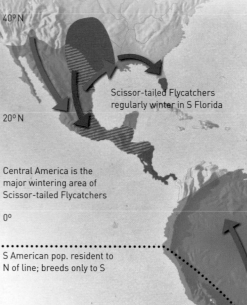

THE MAJORITY OF MIGRANT VERMILION FLYCATCHERS (below) move south well into Mexico or beyond for the winter. At least a few, however, regularly travel east to winter along the northern Gulf coast and as far inland as the Mississippi delta.

Most South American Vermilion Flycatchers are resident, although those that breed in the south of the continent migrate north for the austral winter (see pp. 10–13).

40° N

20° N

Scissor-tailed Flycatchers regularly winter in S Florida

Central America is the major wintering area of Scissor-tailed Flycatchers

0°

S American pop. resident to N of line; breeds only to S

20° S

MIGRATION
BREEDING
JAN FEB MAR APR MAY JUN JUL AUG SEP OCT NOV DEC

SCISSOR-TAILED FLYCATCHER
Tyrannus forficatus
Wingspan
14–15½ in / 36–39 cm
Weight
1½ oz / 43 g
Journey length
1,2500–2,500 miles /
2000–4,000 km

Thrushes

Robins arrive right on time.

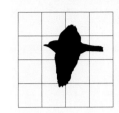

AMERICAN ROBIN
Turdus migratorius
Wingspan
14½–16½ in / 37–42 cm
Weight
2¾ oz / 77 g
Journey length
0–4,000 miles /
0–6,400 km

More than any other North American bird, the American Robin symbolizes the arrival of spring, largely perhaps because of the predictable timing of its spring migration. On its northward journey, this thrush is a relatively slow migrant, following the 37°F (2°C) isotherm to appear in northern regions immediately after the spring thaw, when average temperatures rise to this level. It may take the birds 11 weeks to travel the 3,000 miles (4,800 km) from Iowa to Alaska. Males migrate earlier than females and arrive in northern states by March, but do not begin singing until the females arrive in April.

Most North American thrushes, including the Veery, and the Swainson's, Hermit, and Gray-cheeked thrushes—as well as the American Robin—are nocturnal migrants. In contrast to the American Robin, the Gray-cheeked Thrush is a relatively fast traveler, only reaching the southern United States in late April and arriving back at the Alaskan breeding grounds within a month.

Varied Thrushes nest from Alaska to northern California and typically winter in western states to the south of their nesting areas. They have also been seen at bird feeders in many eastern states in winter. Most thrushes forage on the ground for insects, snails, and earthworms. The Hermit Thrush, however, includes more fruit in its diet than other species and commonly winters in the southeastern states.

AMERICAN ROBINS breed over most of North America south of the tree line. They are especially common in the urban centers of the east, although they do not breed in coastal areas from Louisiana to New York, including most of peninsular Florida. (There is an isolated resident population in Tampa.) American Robins winter from southern Canada to Guatemala, Cuba, and the Bahamas.

They are nocturnal migrants, traveling in small groups of well-spaced birds. Feeding stops on the fall migration are often at hackberry, sugarberry, or other fall-fruiting trees. In spring the birds halt to replenish reserves in grassy areas.

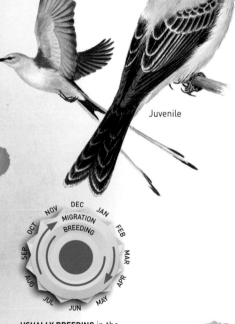

Scissor-tailed Flycatcher

Juvenile

USUALLY BREEDING in the southern Great Plains and wintering in Mexico and Central America, the Scissor-tailed Flycatcher (above) also regularly winters in southern Florida. On migration it wanders widely, even to southern Canada.

Empty nests outside the Great Plains, in eastern Mississippi, for example, are common and suggest that birds are paired when they begin the spring migration. The return of birds to areas outside their normal range may indicate fidelity to routes and sites of successful nesting.

American Robin

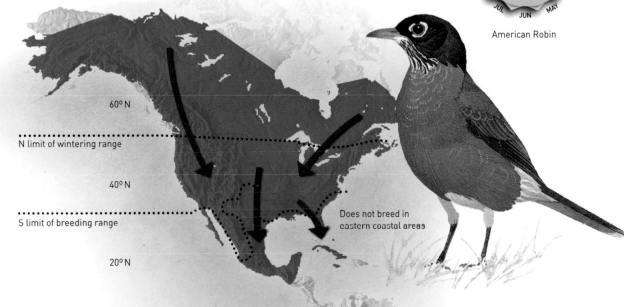

60° N

N limit of wintering range

40° N

S limit of breeding range

Does not breed in
eastern coastal areas

20° N

Vireos

Cross-continent routes betray ancient distributions.

Uniquely American, the vireos are a group of some 46 foliage-gleaning forest birds with uncertain family connections. Vireo species are about equally divided between North and South America, with one, the Black-whiskered Vireo, largely restricted to the West Indies. Several vireos make long migratory journeys; all are nocturnal migrants.

North American vireos can be divided into species that breed in either eastern or western North America and have similarly divided eastern and western wintering areas. Black-capped, Bell's, and Gray vireos, for example, are generally western breeders that winter in northwestern Mexico. The White-eyed, Yellow-throated, and Philadelphia vireos are eastern species with migration routes through the Greater Antilles, along the east coast of Mexico, or across the Gulf to Yucatán.

Some species' migrations, however, seem to betray the manner in which ranges have changed and expanded over the generations. All populations of the Red-eyed Vireo, which is an eastern species that has spread as far as the Pacific Northwest, migrate to South America via the Gulf States. By contrast, all breeding populations of the Warbling Vireo, a western species that has extended east, winter from western Mexico into southwestern South America. A complex east–west breeding and wintering distribution is shown by the Solitary Vireo, eastern populations of which sometimes winter in the Gulf States, while western populations more frequently winter in Mexico and Central America.

The dynamic nature of species' breeding ranges and the role of migration in range expansion may be indicated by recent northward expansion of the breeding range of White-eyed Vireos. Strong favorable spring winds over the Gulf of Mexico during migration may have caused such migrants to "overshoot" their usual destinations.

Two factors contribute to the permanent expansion of a breeding range: the movement of large numbers of birds and either an improvement in climate or the adaptation by birds to the prevailing climatic conditions. Vagrant Black-whiskered Vireos from the West Indies are occasionally found in Alabama and Mississippi in spring, when there are strong southerly winds. But, since the numbers of this species reaching the Gulf coast are very small, there is no evidence of Black-whiskered Vireos breeding, by contrast to their White-eyed relatives.

RED-EYED VIREO
Vireo olivaceus
Wingspan
9½–10½ in / 24–27 cm
Weight
½–¾ oz / 14–24 g
Journey length
2,500–6,000 miles /
4,000–10,000 km

60° N

40° N

PERHAPS THE MOST COMMON BIRD of eastern deciduous forests, the Red-eyed Vireo is most visible on migration in eastern Mexico. The journey from eastern North America to Central and South America, however, regularly takes some birds through the Bahamas, Cuba, and Jamaica, although rarely as far east as Puerto Rico.

Southern South American breeders are also migratory, wintering in Amazonia.

20° N

0°

Limits of wintering range of Scarlet Tanagers not known

20° S

S American Red-eyed Vireos to N of this line are resident; birds breeding to S migrate N

Red-eyed Vireo

Tanagers

Tree cutting puts pressure on forest nesters.

Confined to the New World, the tanagers breed in South and Central America and the West Indies. Because they are largely fruit-eaters, most are limited to the tropics, and North American breeders usually migrate to winter in warmer latitudes. However, some tropical species and populations also migrate. These movements, such as those of the Black-and-white Tanager of northwestern Peru, are a response to seasonal rainfall cycles and consequent limited availability of food.

The Scarlet and Summer tanagers of eastern North America are birds that favor extensive forest and have no doubt declined as the forests have been altered and fragmented. Summer Tanagers are long-distance migrants, breeding almost into southern Canada and wintering as far south as Brazil and Bolivia. Western Tanagers are also forest birds, traditionally nesting in the open coniferous and mixed woodlands of western North America and wintering from Mexico south as far as Costa Rica.

SCARLET TANAGER
Piranga olivacea
Wingspan
11–12 in / 28–30 cm
Weight
1 oz / 29 g
Journey length
600–4,350 miles /
1,000–7,000 km

SCARLET TANAGERS nest in the northern U.S. and southern Canada and migrate to South America, wintering from Colombia to Bolivia.

This species is a trans-Gulf/trans-Caribbean migrant. Some individuals leave from the northern Gulf coast and fly to the Yucatán peninsula; some fly directly to Honduras and Costa Rica en route to wintering areas in northwestern South America. Others reach the wintering areas via the Greater and Lesser Antilles, although they rarely reach as far east as Puerto Rico on migration. Unlike more southerly breeding tanagers, Scarlet Tanagers feed on bees, wasps, beetles, moths, and their larvae, in addition to a variety of fruits.

✖ DEFORESTATION

More than 80 percent of the forest present in the lower Mississippi valley 250 years ago has been cleared; what remains has been dramatically altered in age and composition, with old, mixed-species forests replaced by young, single-species plantations. There have been similar losses throughout eastern North America as a result of cutting, wetland drainage, and the construction of dams. At the same time, Dutch elm disease and chestnut blight reduced tree numbers as efficiently as the chainsaws.

All species that breed in mature forests have been affected to a greater or lesser extent. Not only have nest sites been destroyed, but the numbers of insects on which these birds feed themselves and their young have also declined. Once numerous, Ivory-billed Woodpeckers, for example, may have been extinct since the 1940s in most of North America. At best, a few may survive in Florida, Mississippi, Alabama, and Cuba. A recent report from Arkansas still needs confirmation.

Diminution of the eastern forest and the ravaging of neotropical forests have more far-reaching effects than these losses to breeding birds. Ecologists are still pondering several questions. To what extent has forest loss changed migration routes for species that used the forests as winter homes? How many species now lack suitable winter habitat? How have the clearing of coastal forests for cities and the deforestation of West Indian islands affected transient migrants? And how many neotropical migrants are limited in their migrations by a lack of suitable habitat along the route?

Foresters argue that the continent has more forest than at any time this century, but for many species that may not be enough.

Main migration of Scarlet Tanagers through W Indies

Most N American Red-eyed Vireos join residents in Amazonia for winter

Molting

♂

♂

Scarlet Tanager

♀

NOV DEC JAN
OCT MIGRATION FEB
SEP BREEDING MAR
AUG APR
JUL MAY
JUN

Wood warblers

Brood parasitism has severely affected one of North America's rarest migrants.

The wood warblers are one of the most significant groups of New World migrants, with 57 species breeding in North America. These birds are primarily insect-eaters, and although many may fly by day over continental North America, most are nocturnal migrants as they cross the Gulf of Mexico. On their northbound journey, the warblers often move only a short distance in the morning and spend much of the day foraging to replenish their energy reserves.

Since they are so dependent on insects, many warblers do not arrive in southeastern states until the leaves on which insects and their larvae feed are well developed on the trees. For this reason it is difficult for southern birders to spot them. As they move northward, they reach the Midwest and northern areas at an earlier stage of leaf development, making them easier to study.

A few warblers supplement their diet with berries and seeds and are thus capable of wintering at more northern latitudes. The Pine Warbler, for example, breeds in eastern North America from southern Canada to southern Florida and the Gulf coast and regularly winters as far north as Arkansas, Tennessee, and Virginia. Similarly, the eastern race of the Yellow-rumped Warbler—the Myrtle Warbler—breeds in the northern coniferous forests of eastern North America and winters regularly through much of the eastern and southeastern United States.

Cape May Warblers are medium-distance migrants, leaving their breeding areas of central and eastern Canada, New York, Maine, and Vermont to winter in the Bahamas and Greater Antilles. Few of these birds reach Central or South America.

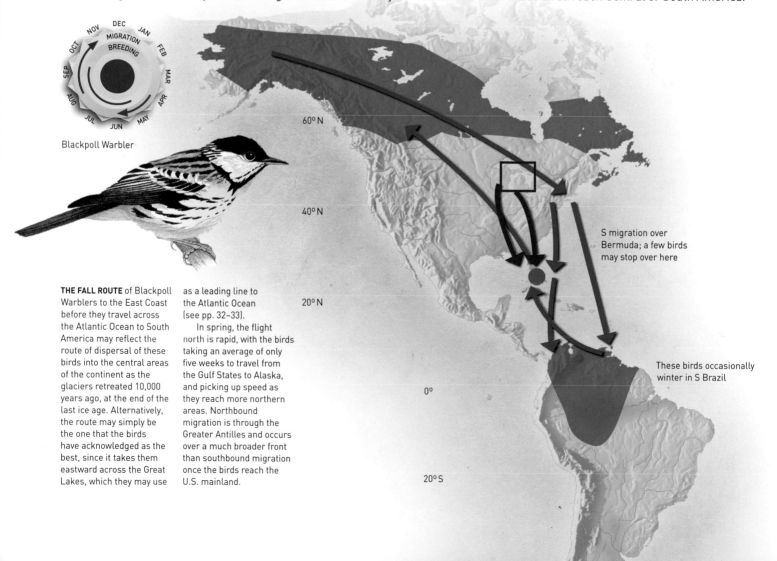

MIGRATION
BREEDING
NOV DEC JAN
OCT FEB
SEP MAR
AUG APR
JUL JUN MAY

Blackpoll Warbler

60° N

40° N

20° N

0°

20° S

S migration over Bermuda; a few birds may stop over here

These birds occasionally winter in S Brazil

THE FALL ROUTE of Blackpoll Warblers to the East Coast before they travel across the Atlantic Ocean to South America may reflect the route of dispersal of these birds into the central areas of the continent as the glaciers retreated 10,000 years ago, at the end of the last ice age. Alternatively, the route may simply be the one that the birds have acknowledged as the best, since it takes them eastward across the Great Lakes, which they may use as a leading line to the Atlantic Ocean (see pp. 32–33).

In spring, the flight north is rapid, with the birds taking an average of only five weeks to travel from the Gulf States to Alaska, and picking up speed as they reach more northern areas. Northbound migration is through the Greater Antilles and occurs over a much broader front than southbound migration once the birds reach the U.S. mainland.

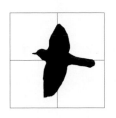

BLACKPOLL WARBLER
Dendroica striata
Wingspan
8–9¾in / 20–24 cm
Weight
⅓oz / 11 g
Journey length
2,500–5,000 miles /
4,000–8,000 km

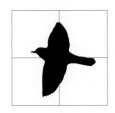

KIRTLAND'S WARBLER
Dendroica kirtlandii
Wingspan
8½ in / 21 cm
Weight
½ oz / 14 g
Journey length
1,200 miles / 1,900 km

Blackpoll Warblers, by contrast, are remarkable long-distance migrants. These common breeding birds of the spruce-fir forests across North America from Alaska to Nova Scotia migrate to northern South America. They follow different routes in spring and the fall, and the fall route in particular has intrigued ornithologists for decades.

Most Blackpolls migrate southeast to the Canadian Maritime provinces and the New England coast. From there it was believed that the birds made a trans-Atlantic flight of more than 2,500 miles (4,000 km) to the coast of South America. To make such a flight, these birds are known to take on energy reserves as fat, increasing in weight from a norm of about ⅓ ounce (11 g) to as much as ⅔ ounce (20 g). While some may make the long flight over the Atlantic, enough records have accumulated to demonstrate that some do fly down the East Coast as far as Cape Hatteras, North Carolina, before striking out for the Bahamas and points farther south. A few seem to continue down the East Coast all the way to Florida.

ON THE CRITICAL LIST

Kirtland's Warbler

Lake Michigan
Lake Huron

The Kirtland's Warbler nests in a restricted area of Huron National Forest, Michigan. One of the bird's few advantages is that it is not an old-growth nesting species, but one that nests in young jack pines. Although the time a jack pine forest is suitable for nesting Kirtland's Warblers is relatively short, it is possible, since the birds' precise requirements are known, to produce good nesting habitat for them. In the Bahamas, where the birds winter, deforestation and monoculture forestry may be causing problems, but knowledge of their preferences in habitat and lifestyle here is limited.

● Known nesting area, 1903–82
▲ Natural jack pine stands

● Known nesting area, 1978–82
● Known nesting area, 1903–82

Kirtland's Warblers have been known for only around 150 years and have always been rare. Their status today is critical. The problems facing them are overwhelming: a small nesting area, cowbird parasitism, a small wintering area, and the hazards of migration.

The invasion of the Brown-headed Cowbird as the forests were fragmented and livestock introduced was nearly fatal. By the 1960s, more than 70 percent of Kirtland's Warbler nests were being parasitized. Removal of cowbirds has reduced this to about 5 percent and the population has responded by increasing more than fivefold in the last 15 years.

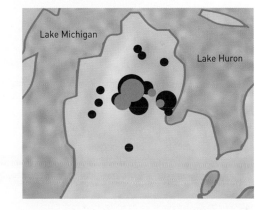

Lake Michigan
Lake Huron

Buntings, grosbeaks, and starlings
Finches move south when snow blankets food stocks.

As seed-eaters, the buntings, grosbeaks, and other New World finches are capable of wintering much farther north than insect-eating birds. Some of these are regular winter visitors to bird feeders well into Canada. Indeed, evidence is mounting that the growing popularity of putting out seed for birds is even beginning to influence their seasonal movements. Evening Grosbeaks, for example, were once considered to be east–west migrants, nesting in the region of the Great Lakes and wintering along the northern Atlantic coast. In recent decades, however, they have appeared, admittedly in irruptive rather than truly migratory fashion, in large numbers throughout the eastern United States.

Although most bird migration typically takes place in fall and spring, in some species it can be renewed in midwinter in response to poor weather or failing food supplies. In periods of excessive snow and cold, American Tree Sparrows may be observed flying near streetlights at night and are often found dead as a result of collisions with radio antenna guy wires, phenomena that normally occur during the birds' nocturnal migration in fall and spring. And the Dark-eyed Junco is commonly known as the "snow bird," since its appearance regularly seems to coincide with the first snowfalls at middle latitudes.

Oregon Junco

Dark-eyed Junco

Slate-colored Junco

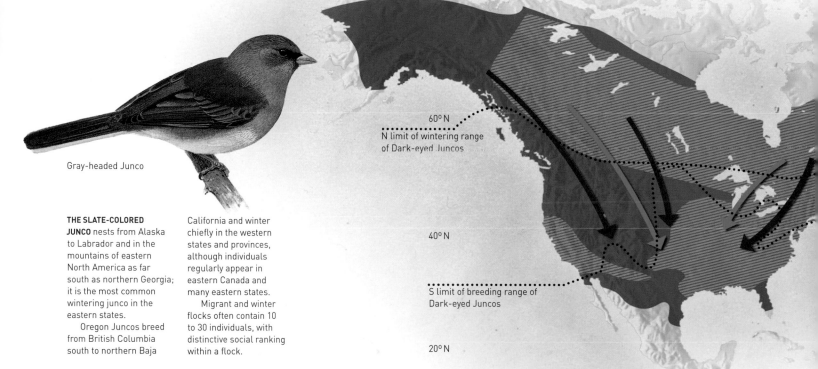

Gray-headed Junco

THE SLATE-COLORED JUNCO nests from Alaska to Labrador and in the mountains of eastern North America as far south as northern Georgia; it is the most common wintering junco in the eastern states.

Oregon Juncos breed from British Columbia south to northern Baja California and winter chiefly in the western states and provinces, although individuals regularly appear in eastern Canada and many eastern states.

Migrant and winter flocks often contain 10 to 30 individuals, with distinctive social ranking within a flock.

60° N
N limit of wintering range of Dark-eyed Juncos

40° N

S limit of breeding range of Dark-eyed Juncos

20° N

DARK-EYED JUNCO
Junco hyemalis
Wingspan
9½–10 in / 24–25 cm
Weight
³/₅–⁹/₁₀ oz / 18–26 g
Journey length
0–2,500 miles / 0–4,000 km

WHITE-THROATED SPARROW
Zonotrichia albicollis
Wingspan
9–10 in / 23–25 cm
Weight
⁹/₁₀ oz / 26 g
Journey length
0–2,800 miles / 0–4,500 km

EUROPEAN STARLING
Sturnus vulgaris
Wingspan
15½ in / 39 cm
Weight
2⁴/₅–3 oz / 80–85 g
Journey length
0–600 miles / 0–1,000 km

A SIMILAR PATH

The European Starling was released in New York's Central Park in 1890 simply because Shakespeare-lover Eugene Scheifflin wanted to introduce to North America all the birds mentioned in Shakespeare's works. Starlings began life in the New World by nesting on the American Museum of Natural History. In less than a century, they had dispersed and were successfully established —often to the detriment of native cavity-nesting birds—from the Atlantic to the Pacific and from Alaska to Mexico.

Migrants in Europe travel from northeast to southwest, and a similar path has been established in eastern North America.

European Starling

Juvenile

There are gender and age differences in the migrations of some finches, including the Dark-eyed Junco. Until 1983 the five races of this species were considered as separate species: the Slate-colored, Oregon, White-winged, Gray-headed, and Guadalupe juncos. The fact that there is some interbreeding among the races has led them to be regarded now as one species. Female Dark-eyed Juncos of all races regularly migrate an average of 300 miles (500 km) farther south than males. Adults also seem to migrate greater distances than juveniles. Juncos are strongly area-faithful, returning each year to the area of their birth to breed.

The behavior of the House Finch clearly demonstrates how ranges expand and migratory behavior changes. These birds were once considered sedentary and, indeed, most populations still do not migrate. In the early 1940s, House Finches from the southern Californian race were introduced to the New York City region. They quickly became established and, in cold winters, shifted their range south and west. One House Finch banded in New York was recovered in North Carolina. And at least some banded birds returned to their northern breeding areas in spring, so they can be considered true migrants. Others that dispersed established new populations so that the species has now expanded throughout eastern North America. Meanwhile, populations in the west have also expanded eastward, taking advantage of bird feeders and other human changes to the landscape such as buildings and reservoirs.

TWO COLOR TYPES, or morphs, of the White-throated Sparrow have been recently identified and studied: those with distinct white striping on the head and those with buff-colored striping. The differences are not fully understood, but it has been suggested that individuals select opposite color morphs as mates.

These birds tend to migrate by the easiest north–south route and do not travel great distances. As seed-eaters and frequent bird-feeder visitors, they can remain as far north as southern Canada throughout the winter. White-throated Sparrows are strongly area-faithful, returning to nest and to winter in the same locations every year.

Although it is not certain whether winter flock members migrate together, once they are all on the wintering grounds, definite dominance hierarchies are established.

White-throated Sparrow

American blackbirds and orioles

Death and disease give huge flocks of these birds a bad name.

It may be the association of the color black with death, or the birds' winter numbers and flocking behavior in open country, or public health warnings of disease associated with large blackbird roosts that give North American "blackbirds" a negative image. Whatever the reason, enormous fall migrant and wintering flocks annually perpetuate the problem.

In North America the name "blackbird" is often misused and misunderstood. Among the "true" blackbirds are the orioles and meadowlarks with which few people find fault; the cowbirds, which as brood parasites are universally unpopular; and such typical blackbirds as the Common, Boat-tailed, and Great-tailed grackles, and the Red-winged, Yellow-headed, Brewer's, and Rusty blackbirds. Certainly, not all black birds are blackbirds: the crows, European Starlings, Lark Buntings, Pileated Woodpeckers, and others belong to several different families.

Loss of and damage to grain in harvest fields and threats to human health are the major causes of concern with migrant blackbirds. Accusations against the birds, which begin when premigratory flocks congregate, build as migrants become more evident, and reach a peak in winter, prior to the birds' dispersal back to the breeding grounds. These problems may have been exacerbated by changing patterns of agriculture and irrigation, which have led to population increases and new migration patterns. Now that irrigation has meant crops can be grown in areas of former desert, Yellow-headed Blackbirds, for example, winter farther north than they did 50 years ago. Against the grain losses, which can be serious for the farmers affected, must be balanced the flocks' destruction of enormous quantities of harmful insects and weed seeds.

The threat to human health derives from the fungal disease histoplasmosis, which is potentially lethal. The fungus causing the disease grows in nitrogen-rich soil, such as that found beneath blackbird roosts. It is, however, dispersed by the wind only when the soil is disturbed, not "carried" by the birds to humans.

Millions of blackbirds have been killed and acres of trees leveled to discourage blackbirds from roosting in specific areas, in the name of histoplasmosis prevention. In fact, clearing vegetation from a roost may create a more serious threat by exposing the nitrogen-rich soils to the wind. The major reasons for concern over blackbird roosts are often the smell and noise, in combination with ignorance of the role these birds play in natural ecosystems.

NEST PIRACY

The Brown-headed Cowbird lays its eggs in the nests of other species, causing their chicks to fail and their populations to decline (see p. 75). Now found in most of North America, this cowbird moves south in winter, flocking by midsummer for migration and often joining flocks of other blackbird species in their winter roosts.

Brown-headed Cowbird

BROWN-HEADED COWBIRD
Molothrus ater
Wingspan
11½–14 in / 29–35 cm
Weight
1⅓–1¾ oz / 39–49 g
Journey length
0–1,250 miles / 0–2,000 km

RED-WINGED BLACKBIRD
Agelaius phoeniceus
Wingspan
12–14½ in / 30–37 cm
Weight
1½–2¼ oz / 41–64 g
Journey length
0–1,550 miles / 0–2,500 km

BOBOLINK
Dolichonyx oryzivorus
Wingspan
10–12½ in / 25–32 cm
Weight
1¼–1⅔ oz / 37–47 g
Journey length
5,000–6,800 miles / 8,000–11,000 km

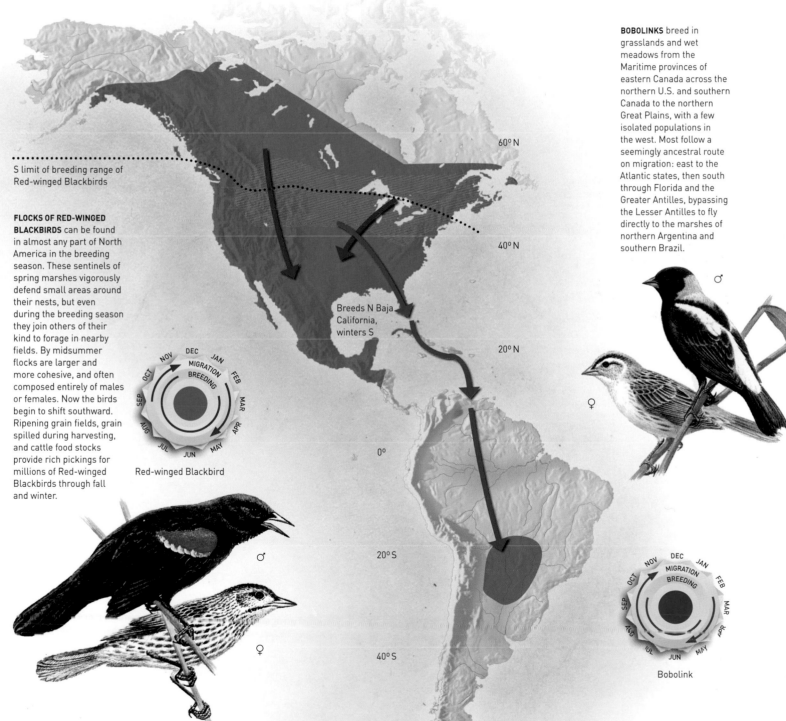

S limit of breeding range of Red-winged Blackbirds

FLOCKS OF RED-WINGED BLACKBIRDS can be found in almost any part of North America in the breeding season. These sentinels of spring marshes vigorously defend small areas around their nests, but even during the breeding season they join others of their kind to forage in nearby fields. By midsummer flocks are larger and more cohesive, and often composed entirely of males or females. Now the birds begin to shift southward. Ripening grain fields, grain spilled during harvesting, and cattle food stocks provide rich pickings for millions of Red-winged Blackbirds through fall and winter.

Red-winged Blackbird

Breeds N Baja California, winters S

BOBOLINKS breed in grasslands and wet meadows from the Maritime provinces of eastern Canada across the northern U.S. and southern Canada to the northern Great Plains, with a few isolated populations in the west. Most follow a seemingly ancestral route on migration: east to the Atlantic states, then south through Florida and the Greater Antilles, bypassing the Lesser Antilles to fly directly to the marshes of northern Argentina and southern Brazil.

Bobolink

60° N
40° N
20° N
0°
20° S
40° S

Eurasian migrants

Long before written records were kept, people in Europe were aware of bird migration. The birds were important as a source of food, and their comings and goings were the practical prompts for agricultural operations such as plowing and sowing. Although they were aware of the birds, those ancient watchers can have had little idea of the extent and complexity of their movements over the Old World and beyond.

In this chapter, the importance of Africa as a reception area for birds summering in Eurasia recurs time after time. A huge number and variety of species, involving an immense total of individuals, leave Europe and Asia for Africa every fall. Best estimates suggest that five billion individuals from more than 200 species regularly migrate to sub-Saharan Africa. But Africa is vast, and the migrant populations are spread over some 7¾ million square miles (20 million km²)—perhaps one migrant per acre, taking into account the birds lost on the journey south. And Africa is not the only destination. The pages that follow detail a complex web of movement within Eurasia, involving both the oceans and other continents.

Eurasian wigeon from Iceland, northern Scandinavia, and north-west Russia make long migratory journeys to winter in north-west Europe, especially Denmark, the Netherlands, Britain, and Ireland.

Patterns of migration

The vast landmass of the continent of Africa dominates Eurasian migration patterns.

Movements of Old World birds are primarily north to south and are dominated, to the south, by Africa. Fewer species use the Indian subcontinent, but many others regularly migrate to the Far East, and a few reach as far as Australia. A major determinant in these patterns has been the way in which the continents have drifted over the last few million years. Africa has always been close to southern Europe and western Asia; India collided with the Asian landmass only 30 million years ago; and Australia has been isolated for some 45 million years.

In Africa and in other southern areas, the prime habitat afforded by the rainforest is not the target area of long-distance migrants. The rainforest lacks seasonality, and many species have evolved to fill every available ecological niche. Long-distance migrants cannot be sufficiently specialist to compete with the residents and are therefore more likely to be found on the fringes. Here, there

✪ HOT SPOTS

These hot spots include islands or headlands where lost small migrants may be seen. Such birds are also funneled through mountain passes, which—with narrow stretches of water—are good sites for soaring migrants. Wetlands are excellent places to watch waterfowl and shorebirds.

- ❶ Fair Isle Observatory, Shetland
- ❷ Cape Clear Observatory, Eire
- ❸ Spurn Observatory, England
- ❹ W Jutland peninsula, Denmark
- ❺ Falsterbo/Ottenby, Sweden
- ❻ Hanko, Finland
- ❼ Rybachiy, Lithuania
- ❽ Texel, the Netherlands/Helgoland, Germany
- ❾ Blakeney Point/Cley Marshes/Minsmere/ Dungeness, England
- ❿ Isles of Scilly, England/Ile d'Ouessant, France

THE OLD WORLD MIGRATION SYSTEMS cover half the globe. They have been molded by the history of the birds involved and modified by their physical environment.

Barriers to migration that are insignificant for one species are insuperable for another. The relatively short crossing of the middle of the Mediterranean holds no terrors for warblers, chats, and other long-distance small migrants, but it is testing for soarers such as storks or broad-winged birds of prey. (In fact, some White Storks and Honey Buzzards do use a middle route made possible by the short distance between Cap Bon in Tunisia and Sicily.)

Mountain ranges are physical barriers that may split routes used by migrants or channel them. Foothills often ensure reliable thermals for soaring birds. Mountainous areas often furnish a variety of habitats that

migrants may find similar to their summer homes. Dotterels from northern Europe, for example, winter in the Atlas Mountains of the Maghreb.

Deserts are not so hospitable. Here, the chances of generalist foragers, such as many long-distance migrants, surviving for any long period of time are remote. Deserts must be skirted, or overflown, as quickly as possible.

The physical problems that birds have to conquer on a route are seldom apparent from a map. A distance of 600 miles (1,000 km) may be a different prospect in spring, when the habitat is productive and welcoming, than in fall when the land is dry and barren. And the northerly winds that helped in the fall may effectively double the distance to be covered in spring. Knowing where you are going may be half the battle, but finishing the journey is far from assured.

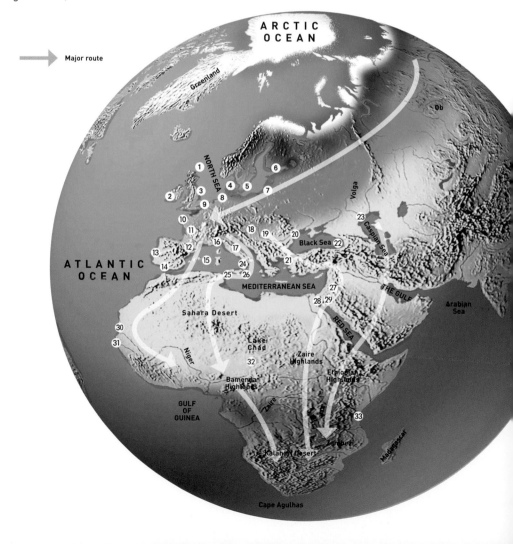

Major route

AROUND FIVE BILLION LAND BIRDS of nearly 200 species breeding in Europe and Asia migrate every year. Some 50,000 birds cross each mile (1.6 km) of Mediterranean coastline every night during the peak fall migration.

THE EURASIAN WRYNECK is unusual among woodpeckers in that it is a long-distance migrant, breeding as far north as northern Scandinavia and traveling to central Africa and southern Asia for the winter.

BIRDS OF THE SIBERIAN RACE of Willow Warbler make an annual migration of up to 7,000 miles (11,300 km) to their winter homes in eastern Africa—a remarkable journey for a bird weighing ⅓ ounce (10 g) or less.

- ⑪ Lac de Grand Lieu, France
- ⑫ Col de l'Organbidexka, France/Spain
- ⑬ Tagus estuary, Portugal
- ⑭ Coto Doñana, Spain/Strait of Gibraltar
- ⑮ Cap Formentor, Majorca
- ⑯ Camargue, France/Col de Bretolet Switzerland
- ⑰ Po marshes, Italy
- ⑱ Lake Neusiedler, Austria
- ⑲ Hortabagy, Hungary
- ⑳ Danube delta, Romania

- ㉑ Strait of Bosporus, Turkey
- ㉒ Black Sea coasts, Turkey/Russia
- ㉓ Volga delta/Caspian Sea, Russia
- ㉔ Capri, Italy
- ㉕ Cap Bon, Tunisia
- ㉖ Straits of Messina, Italy
- ㉗ Azraq, Jordan
- ㉘ Suez, Egypt
- ㉙ Eilat, Israel
- ㉚ Banc d'Arguin, Mauritania
- ㉛ Senegal delta, Senegal

- ㉜ Lake Chad, Nigeria/Chad
- ㉝ Ngulia Lodge, Kenya
- ㉞ Lahore, Pakistan
- ㉟ Royal Chitwan NP, Nepal
- ㊱ L. Ozero Khanka, China/Russia
- ㊲ Shiretoko NP, Japan
- ㊳ Shinhama Waterfowl Preserve, Japan
- ㊴ Yangtze estuary, China
- ㊵ Mai Po marshes, Hong Kong
- ㊶ Dalton Pass, Philippines
- ㊷ Fraser's Hill, Malaysia

are seasonal differences—not winter and summer, but rainy and dry—and the local birds are more generalist and less specialist. Migrants are opportunists, able to exploit such resources as locust swarms or the insects that lose their homes when tracts of land are burned. They are also common in areas where habitats have been modified by human agricultural practices.

Long-distance migrations are by no means confined to direct north–south routes, and birds wintering in southern Africa, for example, may have traveled from the far east of Asia. The routes of enormous numbers of birds take them across the continent from east to west (and sometimes from west to east). This is because the climate in the center of the larger landmasses such as Eurasia is more extreme than on the fringes. This is particularly so for the western part of Europe, which is warmed by the Gulf Stream. Many shorebird and waterfowl species simply move to areas where the winters are ice free, which may be to the west of their breeding ranges, but not far, if at all, to the south. Thrushes and some finches are among the birds that undertake such movements.

Migration is not confined to land birds. There are, in addition, long-distance movements by sea birds across and around the oceans that fringe Eurasia (see pp. 150–51).

Ducks

"First come, first served" means that female Pochards make longer migrations than males.

Large numbers of ducks of many species breed in the northern parts of Europe and Asia and migrate south to warmer areas for the winter. They include representatives of the three major types of ducks: dabbling ducks, such as the Eurasian Teal, Wigeon, and Pintail; diving ducks, including the Tufted Duck, Common Pochard, and Greater Scaup; and sea ducks, among them Goldeneye, Smew, and Common Merganser.

The migratory pattern of the Eurasian Teal can vary from season to season, depending on the weather and habitat conditions that the birds find. Areas such as the Netherlands, Britain, and northern France—the major wintering areas for Scandinavian and western Russian breeders—are generally mild enough for these birds to stay throughout the winter.

Every few years, however, when ice and snow hit, Eurasian Teals are among the first to leave, heading south and west in large numbers to southern France, Spain, and Portugal. A few even cross the Mediterranean to northern Africa. (The closely related North American Green-winged Teal behaves in a similar manner, in mild weather wintering on the prairies; if temperatures fall, continuing southwards, sometimes as far as Mexico.)

Springtime drought on the breeding grounds, drying the pools and marshes, also severely affects water birds such as teals. If they arrive back in western Russia to a spring drought, they move on to look for wetter areas, even though these may be outside their normal breeding range. Equally, drought in late summer forces them off the breeding grounds early, so that migrants appear back in western Europe well before their usual time.

In common with other diving ducks, Pochards obtain almost all their food underwater, diving to depths of 7–13 feet (2–4 m) to reach submerged water plants. In both summer and winter, therefore, they frequent freshwater lakes and reservoirs with extensive shallows where vegetation grows. There has been a considerable increase in Pochard numbers in the western half of the range over the last 100 years, much of which has been attributed to the proliferation of artificial waters. Reservoirs, gravel pits, and ponds all provide Pochards with ideal conditions.

Male Pochards begin their fall flight south with a molt migration (see pp. 24–25), after which they continue their journey, reaching the wintering grounds some weeks before the females and young. The latter arrive to find the first suitable areas completely occupied and must fly on farther south. There is, therefore, a

THE EURASIAN TEAL, at some 14 inches (36 cm) from beak to tail, is one of the smallest dabbling ducks, feeding from the surface of shallow lakes and marshes on seeds, insects, and small invertebrates.

It is at risk during both winter and summer: when waters freeze or marshes dry out, teals face serious problems if they do not respond quickly by moving in search of more favorable conditions. Larger species and those with more varied diets, by contrast, can often wait for conditions to improve.

Green-winged Teal
Anas carolinensis

Eurasian Teal
Anas crecca

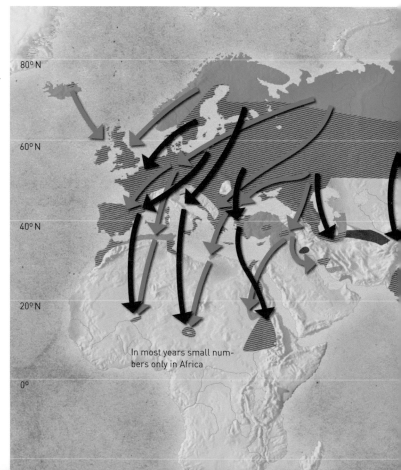

In most years small numbers only in Africa

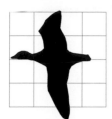

EURASIAN TEAL
Anas crecca
Wingspan
23–25 in / 58–64 cm
Weight
9–16 oz / 250–450 g
Journey length
300–3,000 miles /
500–5,000 km

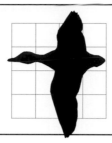

COMMON POCHARD
Aythya ferina
Wingspan
28–32 in / 72–82 cm
Weight 1 lb 9 oz–2 lb 7 oz /
700–1,100 g
Journey length
190–4,600 miles /
300–7,500 km

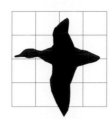

SMEW
Mergus albellus
Wingspan
22–27 in / 55–69 cm
Weight 1 lb 3 oz–1 lb 12 oz /
550–800 g
Journey length
600–2,800 miles /
1,000–4,500 km

AS SOON AS THEIR PART
in the breeding cycle is
complete and the females
have started to incubate
a clutch of eggs, male
Pochards leave the breeding
grounds. Substantial
flocks of many thousands
gather on large lakes—
such as the Ijsselmeer
in the Netherlands and
Ismaninger reservoir,
Bavaria—to undergo
their annual molt
before proceeding to
the winter quarters.

Common Pochard

IN WINTER, Smews are
widespread on fresh and
brackish water. In the
breeding season, however,
they are more restricted
by the need to find holes
in trees in which to nest,
not too far from water.
Areas of mature forests
dotted with lakes and river
banks where old woodland
reaches the water are
favorite haunts. Many
pairs nest in close
proximity, perhaps
making use of a stand
of dead trees containing
plenty of holes.

pronounced segregation of the genders during the winter, with
males predominating in the north of the range and females in
the south. This widespread phenomenon is also known to occur
in other species of ducks, including the Goldeneye, Wigeon, and
Black Scoter.

Smews breed in a broad swathe across northern Europe and
Asia from Sweden almost to the Pacific Ocean. Some 15,000 winter
in northwestern Europe, with more farther east: at least 25,000
have been counted in the Sea of Azov, in the northeastern corner
of the Black Sea, and nearly as many around the Volga delta in the
north Caspian.

The majority of Smews migrating to northwestern Europe spend
the fall in the Baltic Sea, moving on only when forced to do so by
freezing conditions. They may not reach the Netherlands, their
ultimate destination, until December or even January, and must
then start back northeast again in March.

Shorebirds

Farming and freak weather conditions can bring death to many shorebirds.

Lapwings and Redshanks are among the most familiar Eurasian shorebirds. Both are widespread, breeding in lowland areas from western Europe eastward through Russia to the Pacific. Both species have declined in the last 50 years, primarily due to summer habitat loss, but also as a result of their susceptibility to harsh winter weather.

Lapwings favor mild, often maritime, areas in winter. Britain, Ireland, and northern France are near the north of the winter range, with Russian and Scandinavian breeders flying south and west to join residents that may move only as far as the nearest coasts from breeding grounds inland. However, cold winters in these areas may force the migrants on to the major wintering sites in the Iberian peninsula and the western Mediterranean.

Redshanks find food in the mud of coastal estuaries in winter and may starve if a sudden drop in temperature coats the mud flats with ice. A single spell of freezing weather in February 1992 killed almost 50 percent of the 4,000 Redshanks wintering on the Wash in eastern England. Counts of dead birds after sudden cold spells always include a high percentage of Redshanks.

The major contributing factor to the loss of breeding habitats for shorebirds is change in agricultural practices. In western Europe in particular, wet meadows have been drained, and grasslands rich in insect life have been plowed and reseeded or turned over to growing crops.

Lapwings nest on the ground in open fields and have always been vulnerable to trampling by cattle and sheep. Farmers who roll their pastures in late March and April to firm up the soil pose a still greater threat, destroying many nests. Lapwings are, however, resilient: they can lay a new clutch one to two weeks after losing the first and may go on to have a third or fourth clutch if previous broods fail. They have also proved capable of recolonizing former breeding grounds, nesting on reflooded previously drained wet meadows, for example.

The Redshank has suffered most as a breeding species from the drainage of low-lying wet fields. Patches of standing water and shallow ditches are essential feeding places for both adults and, especially, young, but are of no use to farmers who want to grow grass and crops. Redshanks can find suitable breeding grounds in coastal areas, but even here they are not secure: an unusually high tide can sweep nests away, and reclamation schemes are destroying this habitat.

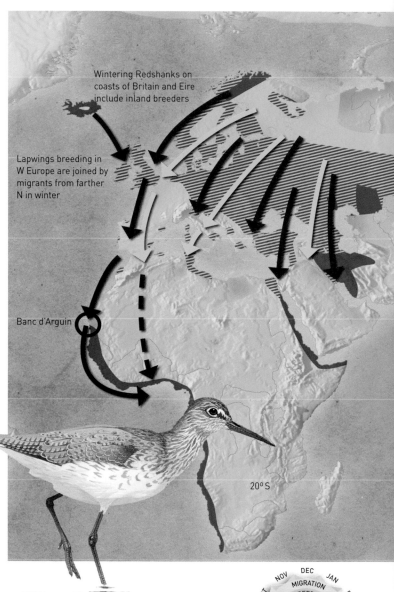

Wintering Redshanks on coasts of Britain and Eire include inland breeders

Lapwings breeding in W Europe are joined by migrants from farther N in winter

Banc d'Arguin

20° S

CLASSIC LEAPFROG MIGRANTS, Redshanks breeding in more northern areas travel greater distances in winter than those that nest farther south. Redshanks breeding in northern Scandinavia winter around the coast of western Africa; those breeding in western Europe venture no farther south than the Mediterranean.

Common Redshank

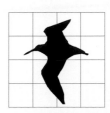

COMMON REDSHANK
Tringa totanus
Wingspan
23–26 in / 59–66 cm
Weight
3–5 oz / 90–150 g
Journey length
300–4,000 miles /
500–6,500 km

NORTHERN LAPWING
Vanellus vanellus
Wingspan
32–34 in / 82–87 cm
Weight
7–10½ oz / 185–300 g
Journey length
300–2,800 miles /
500–4,500 km

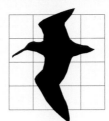

**SLENDER-BILLED
CURLEW**
Numenius tenuirostris
Wingspan
31–36 in / 80–92 cm
Weight
10½–21 oz / 300–600 g
Journey length 2,500–3,700
miles / 4,000–6,000 km

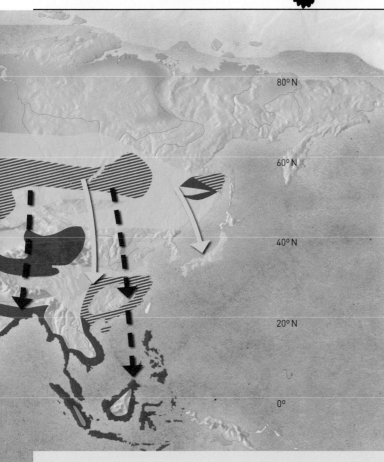

Northern Lapwing

LAPWINGS ARE COMMON,
despite recent declines:
some 100,000 pairs
nest in the Netherlands
and similar numbers in
Germany, with perhaps
twice that many in Britain.
This species is on the move
in every month of the year.
Immediately after the
breeding season, which can
end as early as May, many
adult birds nesting in
Europe and western Asia
start the journey westward.
This dispersal continues
through the summer until
it merges into the fall
migration, when the young
birds head west and south
to the wintering grounds.

WITNESSES TO A GRADUAL EXTINCTION?

In the last 100 years,
Slender-billed Curlew
numbers have fallen
dramatically. Whether
this is due to loss of
breeding or wintering
habitat or to being shot
on migration is uncertain.
In addition, its similarity
to the Eurasian Curlew
may have distorted
knowledge of its former
range and numbers.
    The probable
breeding grounds are
in Siberia and the birds
migrate across Hungary,
Greece, Romania, and
Turkey to winter in
Tunisia, Algeria, and
Morocco. Up to 900
were seen in Morocco
in 1964, but no more
than 100 in the 1980s
and, more recently, only
ones and twos, though
intriguingly one in northeast England in 1998. It
seems very probable
that the species, if not
quite extinct, very soon
will be.

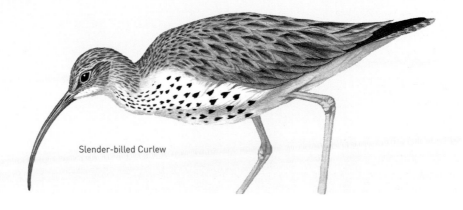

Slender-billed Curlew

Storks

The traditional bringer of babies is threatened at each stage of its migratory journey.

Living alongside human beings has taken its toll on the White Stork. Changing fashions in architecture have dramatically reduced the number of suitable nesting places on buildings, while—from the 1950s onwards—the increased use of agrochemicals severely affected the populations of insects, worms, and frogs on which these birds rely for food. It is estimated that the European population of White Storks had declined by up to 50 percent overall (up to 90 percent in some countries) by the late 1980s. Since then, however, there has been a welcome increase in the majority of countries, brought about by increased protection (although some shooting on migration still occurs), provision of artificial nesting sites, and the banning of pesticides.

The White Stork is a soaring bird, relying on thermals to gain height, then gliding (and sometimes slowly flapping) to the next thermal (see pp. 16–17). Like other thermal soarers, White Storks avoid long sea crossings where possible. This restricts those migrating to Africa to one of two routes over the Mediterranean Sea: across the Strait of Gibraltar in the west or the Bosporus, Turkey, in the east. Broadly speaking, birds breeding west of a line running from Switzerland to the eastern Netherlands head for Gibraltar, those to the east of the line (by far the majority) fly over Turkey.

The reliance on warm-air thermals restricts White Storks to daytime migrations, and consequently they become concentrated at these narrow crossing places. These large, highly visible

THE EUROPEAN BLACK STORK population is small: most countries shelter only a few hundred pairs. Numbers have declined over the last century as the wetlands on which they feed have been drained.

Black Storks are capable of sustained flight and often choose the long sea crossings that true soarers avoid. Some cross into Africa over the Tunisian coast; others have been seen "island hopping" in Sicily, Crete, and Cyprus.

German and Danish breeders are thought to favor crossing the Mediterranean at Gibraltar, although of a brood of four banded in Denmark, two were picked up on course in northern France and the Netherlands, while two had headed southeast and were found in Hungary and Romania.

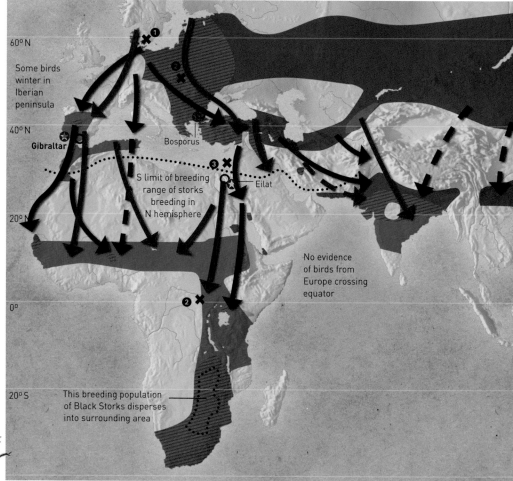

Black Stork

BLACK STORK
Ciconia nigra
Wingspan
4 ft 9 in–5 ft 1 in / 1.45–1.55 m
Weight 5 lb 8 oz–7 lb 11 oz /
2.5–3.5 kg
Journey length
1,250–4,000 miles /
2,000–6,500 km

WHITE STORK
Ciconia ciconia
Wingspan
5 ft 1 in–5 ft 5 in / 1.55–1.65 m
Weight 5 lb 1 oz–9 lb 11 oz /
2.3–4.4 kg
Journey length
1,250–6,500 miles /
2,000–10,500 km

flocks spiral round and round, gradually climbing until they have gained enough height to glide across the sea. Flocks of 10,000 are not unusual at the Bosporus in the fall. Some 340,000 have been counted during the two-month fall migration period. A further 35,000 White Storks each year cross the Mediterranean at Gibraltar.

Once they are safely in Africa, the birds may face a journey of several thousand miles before they reach their winter homes south of the Sahara. The warmer weather as they move farther south, however, means that thermals become increasingly common and they can make good progress. Flying between 3,300 and 8,200 feet (1,000 and 2,500 m) above the ground, they can attain speeds of around 28 mph (45 km/h).

Although more common, the thermals on this leg of the journey still have to be located. White Storks maximize their chances by flying in large, widely spread flocks, typically up to 500 birds across a front some 1,600 feet (500 m) wide. As soon as some of the birds in the flock find the thermal, they cease their forward movement and start to spiral. The rest of the flock soon follows suit.

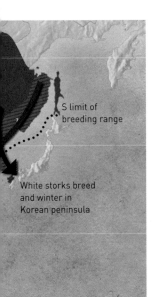

S limit of
breeding range

White storks breed
and winter in
Korean peninsula

CENTURIES AGO WHITE STORKS discovered that chimneys, roofs, towers, and haystacks offered safe and secure nesting places. In many parts of central and northern Europe (the nest, right, is in Austria), these man-made structures came to be preferred to the more common nest site in the branches of a tree. This proximity to people has given storks a special place in human affections: in many countries, the myth exists of the White Stork as the bringer of babies, a tradition that equates the return of migrant birds in spring with fecundity.

White Stork

✖ THREATS

❶ LACK OF NESTING SITES Replacement of traditional housing stock by concrete skyscrapers, particularly in Denmark and northern Germany, has severely restricted available nest sites.

❷ HABITAT DEGRADATION In Eastern Europe due to increasing use of agrochemicals; in Africa due to drainage and use of DDT.

❸ HUNTING In the Middle East.

Birds of prey ①

Finding the shortest possible sea crossing is the prime requirement of soaring birds of prey.

The migrations of Honey-buzzards are typical of Eurasian soaring birds of prey. The majority head for one of three routes: via Gibraltar, through the Bosporus, or over eastern Turkey. These three routes are used both in fall and in spring. Birds have also been sighted using a fourth, over Eilat, Israel, at the head of the Red Sea, where the Dead Sea valley forms a perfect leading line (see pp. 32–33), on the spring migration north only.

Honey-buzzards breed in forests and open woodland right across Europe and into western Russia. Populations increase from west to east: there are, for example, only a handful of pairs in Britain. Regardless of breeding area, however, all Honey-buzzards winter in sub-Saharan Africa. Generally, birds breeding west of a line from Sweden to central Europe take the Gibraltar route, while those breeding to the east head for the Bosporus. Russian breeders hug the Black Sea coast before turning toward Africa. A small number make the long Mediterranean crossing, passing into Africa over the coasts of Algeria and Tunisia and probably heading straight over the Sahara. There is also a significant return in spring via this route: 8,000 individuals were counted in Tunisia in the first half of May 1975.

Falsterbo

Ospreys occasionally winter on Mediterranean islands

Route not known

FUNNELS AND FLYWAYS

Broad-winged birds of prey, such as eagles, buzzards, and harriers, migrate by soaring on thermals. Since thermals occur only over land, these birds avoid crossing water where possible.

Birds migrating from Europe to Africa must cross the Mediterranean and are funneled into places that give them the shortest route over water: via the Strait of Gibraltar in the west; over the Dardanelles and the Bosporus, then the Red Sea in the east; or through Italy, Sicily, and Tunisia.

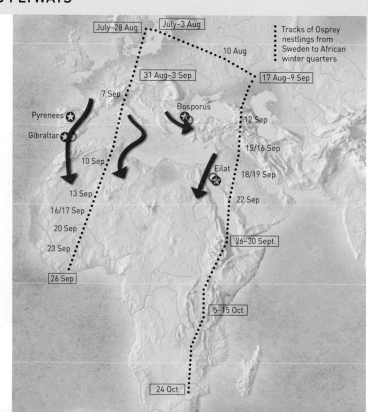

July–28 Aug   July–3 Aug

10 Aug

Tracks of Osprey nestlings from Sweden to African winter quarters

31 Aug–3 Sep   17 Aug–9 Sep

7 Sep

Bosporus

Pyrenees

12 Sep

Gibraltar

15/16 Sep

10 Sep

Eilat

18/19 Sep

13 Sep

22 Sep

16/17 Sep

20 Sep

23 Sep

26–30 Sept

26 Sep

5–15 Oct

24 Oct

Honey-buzzard

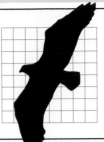

EUROPEAN HONEY-BUZZARD
Pernis apivorus
Wingspan
4 ft 5 in–4 ft 11 in / 1.35–1.5 m
Weight
1 lb–2 lb 3 oz / 450–1,000 g
Journey length
2,500–6,000 miles /
4,000–10,000 km

OSPREY
Pandion haliaetus
Wingspan
4 ft 9 in–5 ft 7 in / 1.45–1.7 m
Weight
2 lb 10 oz–4 lb 7 oz / 1.2–2 kg
Journey length
2,500–6,000 miles /
4,000–10,000 km

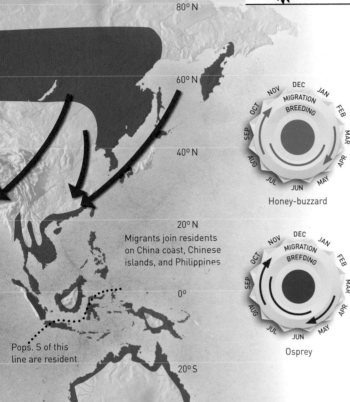

Honey-buzzard

Migrants join residents on China coast, Chinese islands, and Philippines

Pops. S of this line are resident

Osprey

These migrations are spectacular affairs. More than 110,000 Honey-buzzards have been counted over Gibraltar during the fall migration, with daily peaks exceeding 13,000. Fewer birds use the Bosporus, with seasonal totals reaching around 25,000, while 140,000 birds have been counted in eastern Turkey.

These dramatic figures are nonetheless eclipsed by the spring migration over Eilat. Six times between 1977 and 1988, more than 350,000 Honey-buzzards were recorded, while in spring 1985, a total of 850,000 (a significant proportion of the entire Eurasian population) used this route to return to their breeding grounds. In that year, a special study was mounted that extended the usual observation zone west into the Negev desert. On May 7, 1985, some 227,800 Honey-buzzards flew over the desert on their way north.

These birds are daytime migrants, setting off from their roosts at sunrise. When they reach Eilat, they may land again and wait for the usual northerly morning winds to start blowing. Although these are head winds, the birds use them to gain height. They also profit from the lift generated as the air warms and thermals start to bubble up.

IN PREPARATION for its journey, a Honey-buzzard puts on fat, averaging a weight increase from 22 ounces (625 g) to 32 ounces (900 g), enough to fly the 4,350 miles (7,000 km) from northern Europe to tropical Africa in one try. These birds are specialist feeders, taking both adults and young from wasps' and bees' nests. Although they will eat fruit and berries when necessary, the ability to migrate without having to search for food en route is an obvious advantage.

✖ THREATS

HUNTING Honey-buzzards are under pressure in Italy, Malta, and Tunisia

OSPREYS ARE STRONG FLIERS, relying on flapping flight and regularly crossing long stretches of water, including the Baltic and Mediterranean seas. They are broad-front migrants and do not funnel into specific routes as other birds of prey do. Satellite radio-tracking has revealed that birds from a single breeding area may winter far apart. The two tracks shown on the map opposite are of nestlings reared in Sweden. One followed a fairly direct route to West Africa, taking just a month to get there, while the other bird first of all headed south-east before flying south to southern Africa, taking nearly three months for the journey.

Birds of prey ②

Breeding late allows Eleonora's Falcons to prey on migrating songbirds.

Eleonora's Falcons are highly specialized birds of prey that nest on the cliff faces of Mediterranean islands in colonies varying in size from a few to 200 pairs. The total population is about 6,000 pairs. In the late fall, the birds head eastward through the Mediterranean to Egypt, where they turn south and make for Madagascar. Recently, small numbers have also wintered in Tanzania.

Most Mediterranean birds nest between March and May, but Eleonora's Falcons delay breeding until late summer, laying their first eggs at the end of July and hatching the first young about a month later. The chicks are in the nest from late August to early October, a period that coincides with the migration of songbirds south from Europe across the Mediterranean and into Africa. It is on these songbirds that the falcons mainly feed.

Each pair of Eleonora's Falcons rears an average of two young a year so that by the end of the breeding season there are at least 20,000 individuals. As many as 10 million songbirds could be taken by this number of falcons. (The total number of songbirds leaving Europe every fall runs into thousands of millions.)

These falcons, like other birds of prey, pluck their victims' feathers before eating them or feeding them to their young. An examination of these feathers, collected after the young have fledged, shows which species the falcons are taking, and in what proportions. More than 90 different species have been identified, with warblers, shrikes, swifts, small thrushes, larks, and flycatchers all common.

Prey remains also reveal a great deal about songbird migration with some species more common than indicated by direct observations. Grasshopper Warblers, for example, are rarely seen on migration, yet regularly fall prey to Eleonora's Falcons nesting off the coast of Morocco, suggesting that this is a major route out of Europe for these birds.

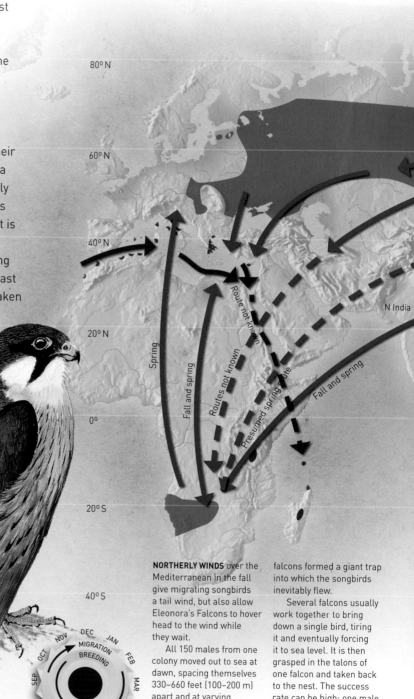

Eleonora's Falcon

NORTHERLY WINDS over the Mediterranean in the fall give migrating songbirds a tail wind, but also allow Eleonora's Falcons to hover head to the wind while they wait.

All 150 males from one colony moved out to sea at dawn, spacing themselves 330–660 feet (100–200 m) apart and at varying heights, up to 3,300 feet (1,000 m) above the water. This "barrier" of waiting falcons formed a giant trap into which the songbirds inevitably flew.

Several falcons usually work together to bring down a single bird, tiring it and eventually forcing it to sea level. It is then grasped in the talons of one falcon and taken back to the nest. The success rate can be high: one male delivered five songbirds to its mate and young within 35 minutes.

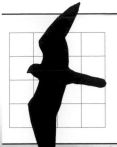

ELEONORA'S FALCON
Falco eleonorae
Wingspan
3 ft 7 in–4 ft 3 in / 1.1–1.3 m
Weight
12–18 oz / 340–500 g
Journey length
3,700–6,000 miles /
6,000–9,500 km

RED-FOOTED FALCON
Falco vespertinus
Wingspan
27–31 in / 66–78 cm
Weight
5–7 oz / 130–200 g
Journey length
4,500–7,500 miles /
7,200–12,000 km

LESSER KESTREL
Falco naumanni
Wingspan
23–29 in / 58–72 cm
Weight
3–7 oz / 90–200 g
Journey length
1,900–6,000 miles /
3,000–9,500 km

Red-footed Falcon

RED-FOOTED FALCONS breed from eastern Europe across central Siberia; a separate race breeds in eastern China, southeastern Russia, Mongolia, and Korea. All winter in the savannas of southeastern Africa.

Most Siberian breeders move west into southern Europe to join the European birds on their journey over the eastern Mediterranean and eastern Sahara. A few take the more direct route south of the Caspian and over Arabia.

Red-footed Falcons breeding in China face an extraordinary migratory journey. In the fall, they fly south of the Himalayas to India. They arrive in November and stay for some weeks, feeding on the abundance of insects to put on fat and awaiting the northeast monsoon. This provides them with a tail wind for the 1,900-mile (3,000-km) flight across the Indian Ocean to Africa.

LESSER KESTRELS (left) nest colonially, groups of 15 to 25, and occasionally as many as 100, pairs breeding in holes in buildings and cliffs. They are also tolerant of humans and often use buildings for their communal winter roosts in South Africa.

These birds are primarily insect-eaters, feeding particularly on grasshoppers, crickets, locusts, and beetles. The increasing use of pesticides on crops has affected populations in many countries of southern Europe, with massive declines in numbers noted.

Gamebirds

These reluctant fliers may make lengthy migrations.

It is difficult to believe, seeing a gamebird in flight, that these skulking birds that take to the air only reluctantly, when flushed, can achieve lengthy migrations, often crossing large expanses of sea or desert.

Gray Partridge migrations are typical of those of many gamebirds, with birds breeding in the milder areas of western Europe being sedentary and more easterly breeding populations moving south to avoid the winter snows. The Common Quail, however, is not only one of the smallest gamebirds but also the most traveled, breeding widely in Europe and into central Asia and wintering in sub-Saharan Africa and in India.

One of the most striking features of Common Quail migration is that the numbers breeding in the north of the range in Europe vary enormously from year to year. Normally, only a few hundred pairs nest in the Netherlands and Britain, but at relatively rare intervals, there are major influxes, multiplying the usual population up to 10 fold. This occurred on six occasions in the last century, the

latest being in 1989, when some 2,500 pairs were estimated to be breeding at the northern extremes of the range instead of the more usual 100 to 300 pairs.

The reasons for these influxes are still unclear. They seem to occur when a warm, dry spring in northern Europe coincides with a period of southeasterly winds. This combination might cause the Quails to overshoot their more southerly, traditional nesting areas and arrive farther north than they intended. Then, since they find the habitat suitable, they stay to breed.

COMMON QUAIL
Coturnix coturnix
Wingspan
13–14 in / 32–35 cm
Weight
2½–5 oz / 75–140 g
Journey length
600–3,400 miles /
1,000–5,500 km

DESPITE REFERENCES to Quail migration in the Old Testament, many aspects of it are still poorly understood. It is known, however, that these birds are subject to intense pressure from hunters when on migration through countries on both sides of the Mediterranean Sea. Many tens, perhaps even hundreds, of thousands are trapped and shot in southern Europe and northern Africa each year.

The extent to which this is affecting the total population is uncertain: Quails have, after all, been hunted for centuries. Yet with so much suitable breeding habitat also being destroyed, the high mortality from hunting is more likely to be having a negative effect.

Common Quail

✖ THREATS

HUNTING AND TRAPPING Quails suffer enormous losses in N Africa, particularly Egypt and Libya.

Occasionally winter in W Europe, N Africa

Broad-front migration over Mediterranean and Sahara

S limit of breeding range

S limit of wintering range of N breeders; birds to S of this line are resident

MIGRATION
BREEDING

Crakes and rails

The Corn Crake's tale is one of inexorable decline.

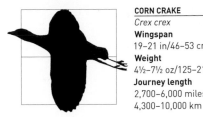

CORN CRAKE
Crex crex
Wingspan
19–21 in/46–53 cm
Weight
4½–7½ oz/125–210 g
Journey length
2,700–6,000 miles/
4,300–10,000 km

One of the most threatened birds breeding in Europe, the Corn Crake has declined drastically in almost every part of its range in the last 30 to 40 years. Figures for Britain are typical: in 1970, the population was estimated at 2,640 pairs; in 1978–79, no more than 750; by 1988, 550 to 600. By 1993 the number had fallen to 475.

Incubating females, and their eggs and young, are killed when the long grass in which they nest is mown for hay or silage. Traditional haymaking involved scything the grass in July or August, a process that rarely harmed the birds. Mechanization, and the change to growing grass for silage instead of for hay, means that the cutting is now done as early as June. In addition, a tractor driver is less likely than a person scything to see a nest, and the practice of mowing in circles from the edges of the field gradually herds any Corn Crake families into the steadily shrinking island in the middle. These birds are extremely reluctant to break cover, so they eventually get caught by the mower blades. If they do run into the open, they are easy prey for crows and gulls.

In an attempt to save this species, British conservation agencies are now sponsoring "Corn Crake–friendly" methods of farming. Since these can mean financial loss to farmers, payments are made for delaying mowing specified fields until August, when the young have fledged. There has been a very welcome effect on numbers, which have increased every year since the measures were introduced, reaching 1,100 pairs by 2005, more than double the low point of 1993.

THE RASPING, far-reaching call of the male Corn Crake (left) signals the presence of a breeding pair. Males call to attract a mate but stop while the eggs are being incubated.

✖ THREATS

CHANGES IN METHODS OF FARMING
Timing of the harvest and mechanization put Corn Crakes at risk in Central Europe and Britain.

Cranes

Observations from a light airplane have revealed many of the wonders of crane migration.

There are references to crane migration in the Old Testament, and it is small wonder that these birds have been part of human consciousness for so long. Huge but graceful, they are noted for their distinctive V-formation flight, their elaborate "dances," and their loud, echoing calls.

Common Cranes breed in marshland across northern Europe and Asia. Most of those breeding in the west of the range, in Sweden and the Baltic countries, migrate southwest to Spain and southern Portugal. Some of these birds, however, together with those from northern Scandinavia and western Russia, migrate southeastward to Turkey, where many stop; others continue either southeast to Iraq and Iran or due south to Egypt, Sudan, and Ethiopia.

Much of what is known about Common Crane migration derives from a study conducted in Sweden one spring using radar to locate the birds, combined with observations from a light airplane. Radar pinpointed the cranes heading north as they crossed the southern coast of Sweden. The aircraft then followed the flocks, with the observers noting what the birds were doing at each stage of the journey.

COMMON CRANES keep to a largely vegetarian diet, taking some animal and insect food, but deriving most of their nourishment from the roots, tubers, leaves, and stems of marsh plants and from crops, including grass and clover.

Migrating flocks must feed en route, and many of their traditional resting places have been used for centuries. Several are now designated refuges so that the cranes are not disturbed as they refuel for the next stage of their journey.

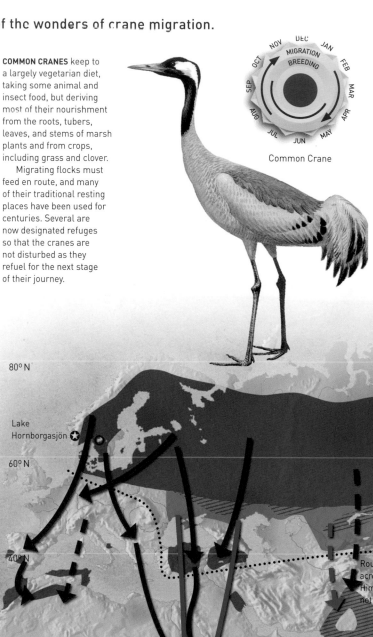

Common Crane

HUMAN INTERVENTION: THE ONLY HOPE?

The Siberian White Crane is seriously endangered. The majority of those remaining, only 2,500 individuals, winter in China where their haunts are at risk from drainage and pollution. Two groups that breed farther west and winter in Iran and India are on the point of extinction; indeed, it is not known whether the Iranian birds still exist. Those that reach India face habitat loss and illegal hunting. In addition, Siberian White Cranes rear few young each summer.

In an attempt to save this species, a captive breeding program has been introduced. Eggs are taken to the United States where the chicks are reared. When they breed, their eggs are returned to Siberia to Common Crane nests. The hope is that the Siberian Crane young will follow the Common Cranes to safer wintering areas farther west.

Siberian White Crane

Lake Hornborgasjön ✪

80° N

60° N

40° N

20° N

0°

Routes across Himalay not kno

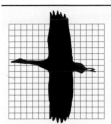

SIBERIAN WHITE CRANE
Grus leucogeranus
Wingspan
7 ft 7 in–8 ft 6 in / 2.3–2.6 m
Weight
11 lb–18 lb 12 oz / 5–8.5 kg
Journey length
2,500–3,000 miles /
4,000–5,000 km

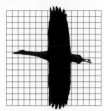

COMMON CRANE
Grus grus
Wingspan
7 ft 3 in–8 ft / 2.2–2.45 m
Weight
8 lb 13 oz–15 lb 7 oz / 4–7 kg
Journey length
1,250–3,750 miles /
2,000–6,000 km

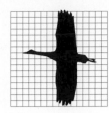

DEMOISELLE CRANE
Anthropoides virgo
Wingspan
5 ft 5 in–6 ft 1 in / 1.65–1.85 m
Weight
4 lb 14 oz–6 lb 10 oz / 2.2–3 kg
Journey length
950–2,800 miles /
1,500–4,500 km

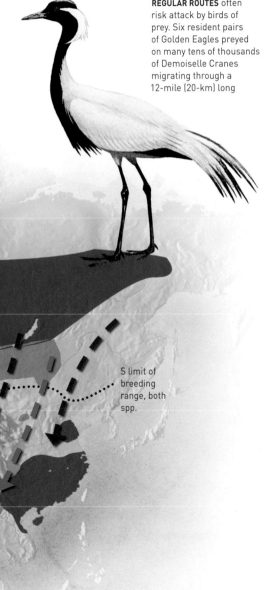

MIGRANTS THAT FOLLOW REGULAR ROUTES often risk attack by birds of prey. Six resident pairs of Golden Eagles preyed on many tens of thousands of Demoiselle Cranes migrating through a 12-mile (20-km) long valley in Nepal. The eagles attacked the lowest-flying cranes, which were up to 3,300 feet (1,000 m) above the valley floor. They were not very successful: only four cranes were killed in 67 observed attacks.

Demoiselle Crane

DURING THE DANCING DISPLAY of a pair of Manchurian Cranes, the birds leap into the air with raised wings, pirouetting, bobbing, and bowing. The dance reinforces the bond between male and female.

S limit of breeding range, both spp.

Common Cranes are thermal soarers, but the tracked birds needed to use flapping flight to cross the Baltic Sea from Germany. When they reached the Swedish coast, they switched back to soaring and gliding. Within a thermal, the flock kept close together, spiraling up to heights of 1,600–6,600 feet (500–2,000 m). At the top of the thermal, the birds peeled off, heading north in a V-formation.

On days when the thermals were weak, the cranes used flapping flight for extended periods, but where possible, they rode the thermals. Because of the time spent soaring in circles to gain height, they averaged 30 mph (50 km/h), less than the 42 mph (67 km/h) they achieved when gliding with occasional flaps between thermals. The energy saved amply compensates for the slower progress.

A similar study of Demoiselle Cranes could yield fascinating results, since most Siberian breeders cross the Himalayas to winter in India. This entails flying at heights of at least 13,000 feet (4,000 m) to clear even the lowest passes. Observations suggest that the birds fly much higher on occasion, perhaps to find favorable tail winds.

Gulls and terns

Young gulls do not share their parents' more limited horizons on migration.

The migratory habits of Eurasian gulls are diverse. Some species are entirely sedentary, others wholly migratory. In addition, there are species that are resident in some areas but leave others for more southerly winter quarters. Almost all the terns are long-distance migrants, traveling great distances every spring and fall.

The closely related Herring and Lesser Black-backed gulls, which often nest together in mixed colonies, behave differently. Herring Gulls are often sedentary, although those that breed in the extreme north of Scandinavia may move south to avoid the most severe winter weather. Lesser Black-backed Gulls, on the other hand, are largely migratory, with the young birds moving much farther in their first fall than the adults, some of which stay at or close to their breeding site throughout the winter.

Young Lesser Black-backs begin to disperse from their nesting areas in Scandinavia and Britain in late July. They are in western France in August, Spain and Portugal in September, and Morocco by October. Some move farther, reaching Mauritania and Senegal. Many adults also move south, but only as far as France and northern Iberia, before heading north again. Some are back in Britain in December and perhaps should be classed as wintering there.

Farther east in the breeding range, fewer birds have been banded, and migration patterns are less clear, although juveniles from northern Scandinavia and northwestern Russia seem to travel farther in winter than the adults. Many juveniles stay in the south throughout their first summer, not returning north until they are 2 years old, although they probably do not breed for another two years.

Black Terns regularly cross the equator on migration. Their journeys begin early, with some dispersal from the breeding grounds in June. The adults move in advance of their young, which often go on random movements (see pp. 28–29), even northward, before they, too, head south. Huge numbers of birds—up to 80,000 have been counted in the IJsselmeer in the Netherlands—assemble in August to feed on concentrations of fish fry. They then head south and west to Iberia, and follow the coastline of Africa all the way to its southern tip. Birds from farther east use the Nile and East African Rift valleys on their southward journeys.

The spring migration of Black Terns is rapid. It does not begin until late March, but birds arrive in western Europe during April and May. Most one-year-olds remain in the winter quarters throughout their first and often second summers, moving north only at the age of 3, when they may make their first breeding attempt. First-time breeders rarely nest in the colonies in which they were reared.

The European breeding range of Black Terns has contracted in the last 50 years, probably due to drainage of the low-lying wetlands in which they nest. Outside the breeding season, however, they have taken advantage of some changes in the landscape, finding plenty of food at sewage plants, reservoirs, and fishponds.

CONSERVATION IN ACTION

The Roseate Tern, threatened in both winter and summer, is the focus of a two-pronged conservation attempt. There are now 1,800 pairs of this species in Europe (1,000 on the Azores, 750 in Britain and Ireland, and 70 in France). This compares with 2,500 pairs in Britain and Ireland, and 600 pairs in France as recently as 1969.

In spring and summer, rats, mink, and gulls prey on Roseate Tern nests on small sand or pebble islands, taking the eggs and chicks. Control of populations of rat and mink has been possible in some areas, and more tall plants are being grown and nest boxes provided to prevent gulls finding the nests.

The winter threat—trapping by children along the coast of western Africa—is being addressed in two ways. Hunting and trapping have been outlawed and an education campaign has been introduced.

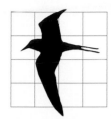

ROSEATE TERN
Sterna dougallii
Wingspan
29–31 in / 72–80 cm
Weight
3–4½ oz / 90–130 g
Journey length
3,000–3,750 miles /
5,000–6,000 km

LESSER BLACK-BACKED GULL
Larus fuscus
Wingspan
4 ft 1 in–5 ft 1 in / 1.25–1.55 m
Weight
1 lb 5 oz–2 lb 3 oz / 600–1,000 g
Journey length
600–4,000 miles /
1,000–6,500 km

BLACK TERN
Chlidonias nigra
Wingspan
25–27 in / 64–68 cm
Weight
2–3 oz / 55–85 g
Journey length
1,850–6,500 miles /
3,000–10,500 km

BETWEEN 1965 AND 1975, the northern limit of the wintering range of Lesser Black-backed Gulls moved north by an average 95 miles (150 km) a year, pushing them into Britain in winter.

An absence of severe winters in those years almost certainly contributed to this shift, but during this period the gulls were also becoming better adapted to living close to humans, nesting successfully on buildings and feeding on garbage dumps. Both habits reduced the incidences of mortality from predation and starvation.

Lesser Black-backed Gull

BLACK TERNS frequent different habitats when breeding, migrating, and wintering. Their diets also differ.

In the breeding season, they are birds of fresh water, nesting beside small pools, shallow lakes, marshes, and swampy meadows. They feed on aquatic insects and their larvae, and on insects caught by hunting low over pastures and rough vegetation.

On the way to their winter homes, they eat freshwater fish, even taking them from backyard ponds. Throughout the winter, they live in coastal waters and feed exclusively on small marine fish.

Black Tern

Small numbers winter
E coast of N America

Elbe estuary

IJsselmeer

Banc d'Arguin

Some this way in spring

A few this way in fall

Some winter as far E
as Gulf of Khambhat

Note: Gulls and terns roost on land but feed at sea, ranging up to 100 miles (160 km) from shore.

Pigeons

These scattered breeders often winter in flocks a million strong.

Most Eurasian pigeons and doves are either sedentary or only partial migrants. They include the homing pigeon, which, although renowned for its navigational skills and used to test theories on how birds migrate long distances with such accuracy, is descended from the sedentary Rock Dove.

Stock Doves and Woodpigeons breeding in the west of the range, in the countries bordering the North Sea and the Mediterranean, are also sedentary, while those nesting farther east journey south and west to avoid the ice and snow of the Eurasian continental winter. The Turtle Dove is an exception, however. These birds breed throughout Europe and in southwestern Asia. All populations are migratory, traveling to sub-Saharan Africa for the winter.

The birds return late in the spring, rarely building their widely scattered nests across woodland, farmland, orchards, and gardens until May. Fall migration may begin in late July but peaks in August and September. Then, huge flocks of European Turtle Doves take one of three major routes over the Mediterranean: across the Strait of Gibraltar (the most popular); via Italy and Sicily; or over the eastern Mediterranean into Egypt. As many as three million Asian breeders have been observed moving southwest over Iraq on a front 60 miles (100 km) wide.

Migrating Turtle Doves are severely threatened by hunting in Mediterranean areas; an estimated minimum of 100,000 are shot annually on the island of Malta, where peak daily passage can reach 20,000 birds.

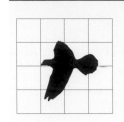

EUROPEAN TURTLE DOVE
Streptopelia turtur
Wingspan
19–21 in / 47–53 cm
Weight
3½–7 oz / 100–200 g
Journey length
600–3,750 miles / 1,000–6,000 km

Egyptian/Sudanese race *rufescens*

Eurasian race *turtur*

ESTIMATED FLOCKS of half a million or more Turtle Doves winter in Africa. These birds feed during the day on harvested rice fields and on the concentrations of seeds left behind after the retreat of floodwaters along some of the larger rivers. Their nighttime roosts are equally huge: one in The Gambia was believed to contain a million birds. Turtle Doves remain south of the Sahara until March, when they begin the journey north.

✖ THREATS

HUNTING
Turtle doves are illegally killed in France, Italy, Malta, and Tunisia.

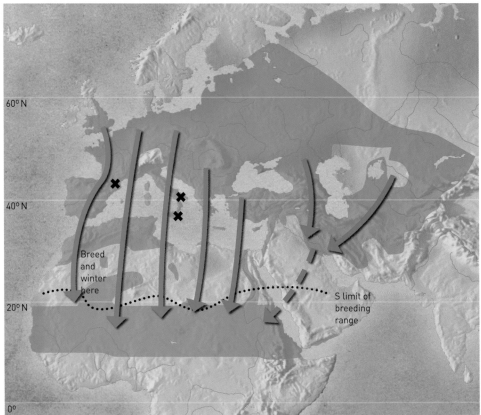

60° N

40° N

20° N

0°

Breed and winter here

S limit of breeding range

Cuckoos

The call of these nest parasites heralds the beginning of spring.

Few birds capture the imagination more than cuckoos. They have a highly distinctive call, often taken to be a harbinger of spring. They lay their eggs in the nests of other birds, forcing surrogates to rear their chicks, which nonetheless inherit their true parents' migration patterns.

Adult Common Cuckoos tend to return to the same breeding area year after year. The female generally defends her territory against other females, but may mate with several males in the course of a season. She searches for suitable foster parents' nests in which to deposit her eggs and usually lays all of them—up to 25 laid at one- or two-day intervals—singly in the nests of the same host species. This has recently been proved to be the same species that reared her. Regardless of species, the foster parents are invariably much smaller than the Common Cuckoo chick that they are obliged to rear.

Once the egg-laying period is over, by the end of July at the latest, the adults are free to migrate and set off southward immediately. There are records of migrants arriving in Africa as early as mid-August. Back on the breeding grounds the young Cuckoo ensures that it gets an adequate food supply by not sharing the nest with any other eggs or nestlings, but evicting them over the side.

About a month after their true parents have left for Africa or Southeast Asia, the young begin their migration. They do not, as might be expected, follow their foster parents. With no knowledge of the route and no parental help, young Cuckoos join the adults on the wintering grounds.

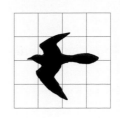

COMMON CUCKOO
Cuculus canorus
Wingspan
22–24 in / 55–60 cm
Weight
3–4½ oz / 90–135 g
Journey length
2,800–7,500 miles / 4,500–12,000 km

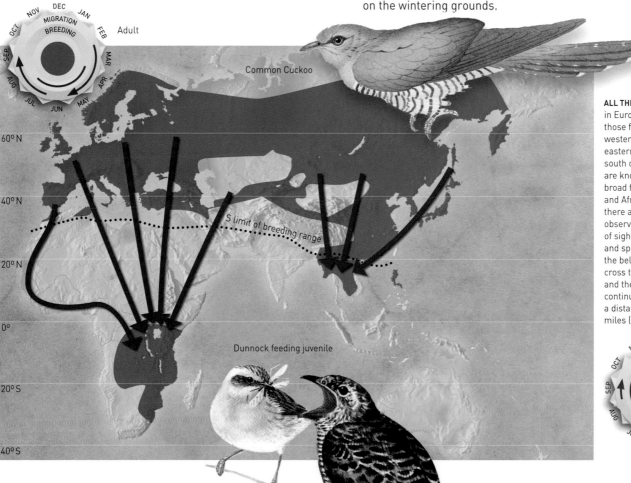

Adult
Common Cuckoo
S limit of breeding range
Dunnock feeding juvenile

ALL THE CUCKOOS breeding in Europe, in addition to those from a large part of western Asia, winter in the eastern areas of Africa, south of the equator. They are known to migrate on a broad front through Europe and Africa, although there are surprisingly few observations. This lack of sightings, in both fall and spring, has prompted the belief that these birds cross the Mediterranean and the Sahara in a single, continuous flight, covering a distance of up to 3,000 miles (5,000 km).

Juvenile

Bee-eaters, rollers, and hoopoes

This exotic trio from southern Europe finds that the Mediterranean is too cold in winter.

Specialist insect-eaters, such as bee-eaters, rollers, and the Eurasian Hoopoe, are restricted to the southern areas of western Europe. Their ranges extend farther north only in eastern Europe and Asia where the continental climate produces hot enough summers to support the necessary abundance of large insects on which they feed. Most birds with such limited diets must migrate south for the winter to areas where insects are still numerous enough.

In common with many birds that fly north to breed in southern Europe, these three species overshoot their breeding areas on occasion and are seen farther north, although they rarely stay to breed. The Eurasian Hoopoe has been proved to breed in Britain at least 30 times, the European Bee-eater four times, and the European Roller not at all.

Although they are restricted to daytime migrations when their food is easy to find, insect-eaters are more fortunate than many migrants in being able to feed on the wing. They have no need to make special, traditional stops, but merely pause on their journey when they come across concentrations of insects. They cope with areas of potentially inhospitable terrain, such as the Sahara and Arabian deserts, by flying non-stop.

All the rollers breeding in Europe and Asia migrate to the eastern half of Africa for the winter. Although they are only thinly distributed, both on the breeding grounds and in winter, total numbers are thought to run into millions. Occasional large flocks occur around good food sources, such as locusts or termites, or when birds are concentrated by storms: on April 13, 1979, some 40,000 to 50,000 European Rollers were seen on migration through Somalia.

With the slow flaps of its broad, striped wings, the Hoopoe in flight looks remarkably like a large butterfly. On the ground, it runs and walks easily in its search for the large insects and occasional small lizards on which it feeds. When it gets excited, it raises its remarkable crest, which completely transforms its shape.

The full extent of Eurasian Hoopoe migration is confused by the presence in its main wintering quarters, in sub-Saharan Africa and in India, of year-round residents. Although these belong to separate races, they are too similar to the European Hoopoe for observers to separate them readily in the field. Thus, observations of movements of feeding flocks in these regions may be of resident or migrant birds (or a combination of the two), and the timing of the migrants' arrivals and departures is difficult to gauge.

EURASIAN HOOPOES (above) bred as far north as southern Sweden a century ago. Since then, the breeding range has retreated south and numbers have fallen, though with some recovery subsequently reversed, about 40 years ago. Although climate change may well have been involved, changes in habitat and, especially, the increased use of pesticides are the more likely causes of recent decline.

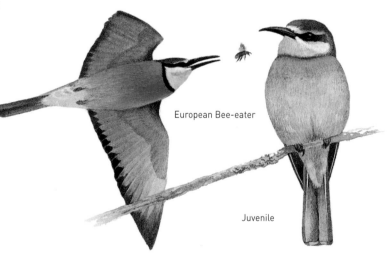

European Bee-eater

Juvenile

BEE-EATERS ARE SPECIALIST FEEDERS, although not as restricted in diet as their name implies. They can be a nuisance to beekeepers, as flocks may gather around groups of hives; a pair can feed its brood of young up to 50 times an hour. However, they also take wild bees and wasps, as well as other flying insects and invertebrates.

To avoid being stung, bee-eaters have a special technique for dealing with bees and wasps. They always take the prey back to a perch, holding it, head forward, in the bill with the insect's head protruding slightly from the bill tip. The bird then bangs the wasp's or bee's head once or twice on the perch before reversing the insect in its bill and rubbing its abdomen repeatedly back and forth against the perch. This tears the sting from the body, or at least discharges its venom.

The bird bangs the bee's head a few more times before swallowing it, head first. The whole procedure takes about 12 seconds.

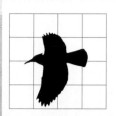

HOOPOE
Upupa epops
Wingspan
16–18 in / 42–46 cm
Weight
2–3 oz / 60–80 g
Journey length
300–3,000 miles /
500–5,000 km

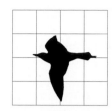

EUROPEAN BEE-EATER
Merops apiaster
Wingspan
17–19 in / 44–49 cm
Weight
1½–2½ oz / 40–65 g
Journey length
1,500–6,500 miles /
2,500–10,500 km

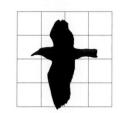

EUROPEAN ROLLER
Coracias garrulus
Wingspan
26–29 in / 66–73 cm
Weight
4–6 oz / 120–176 g
Journey length
1,500–6,000 miles /
2,500–10,000 km

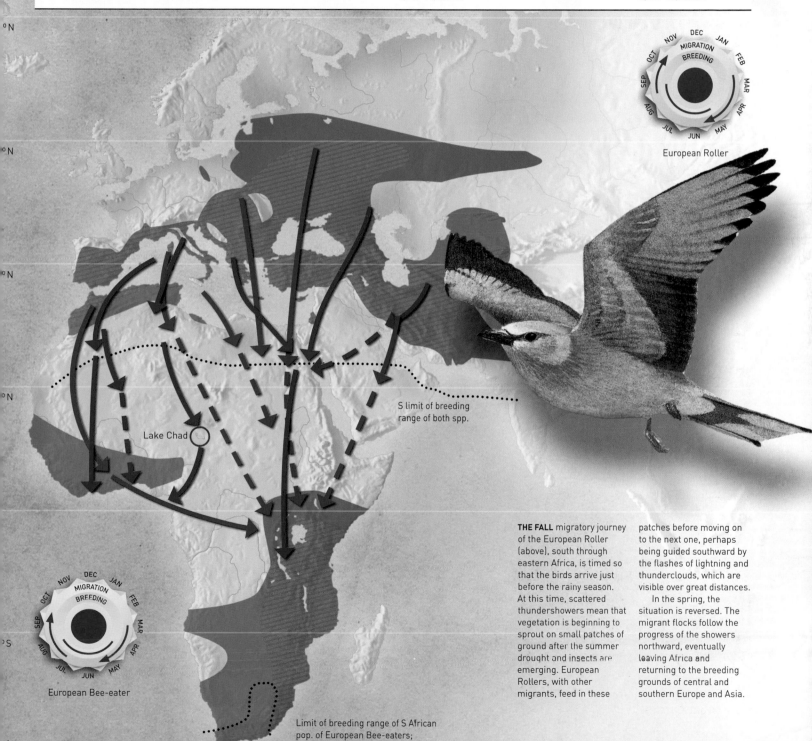

European Roller

S limit of breeding
range of both spp.

Lake Chad

European Bee-eater

Limit of breeding range of S African
pop. of European Bee-eaters;
presumed to disperse N

THE FALL migratory journey of the European Roller (above), south through eastern Africa, is timed so that the birds arrive just before the rainy season. At this time, scattered thundershowers mean that vegetation is beginning to sprout on small patches of ground after the summer drought and insects are emerging. European Rollers, with other migrants, feed in these patches before moving on to the next one, perhaps being guided southward by the flashes of lightning and thunderclouds, which are visible over great distances.

In the spring, the situation is reversed. The migrant flocks follow the progress of the showers northward, eventually leaving Africa and returning to the breeding grounds of central and southern Europe and Asia.

Swifts, swallows, and martins

Slowly but surely is the most effective strategy for migratory young swallows.

Of all northern birds, those that feed exclusively on aerial insects are obvious migrants—forced to leave by the complete lack of food during winter. The Common Swift is the most spectacular of these birds since it is often present in Eurasia for only a quarter of the year, from early May to the first week of August. Others, such as the Sand Martins, may be on the breeding grounds for up to six months, from the end of March to early October.

These birds are completely at home in the air. They move largely by day and, since they feed in flight, could be supposed to migrate in a leisurely fashion. This is indeed the case with swallows and martins, but swifts are remarkably fast migrants. One juvenile found in Madrid, Spain, had been banded only three days earlier in Oxford, England, 780 miles (1,250 km) away. This speed is probably related to their ability to sleep on the wing at night.

Since they are such slow migrants, Sand Martins and Barn Swallows build up tremendous roosts on their southward passage. Over a period of many weeks, between a few hundred and several hundred thousand birds gather in reedbeds near water. Young birds on migration in the fall may spend up to two weeks in an area before moving on, traveling mainly southward, between 60 and 190 miles (100 and 300 km) at a time, thus taking a leisurely couple of months through northern and central Europe to the Mediterranean. Since experienced adult birds know where they are going, they move more briskly, staying in each roost for only a few days.

The northbound flight in spring is faster, at around five weeks, but again the adult birds lead the way. In the wintering area, whether it is in the Senegal River delta for Sand Martins or the gold-mining areas of the Transvaal for Barn Swallows, nighttime roosts may be huge. Again, these are generally in wet reedbeds since they are safe havens against ground-based predators.

Swift and House Martin roosts have not been found, although House Martins often roost in colonies in unoccupied nests. In bad weather, on migration, they have on rare occasions roosted in trees and, if it is exceptionally cold, they may roost on buildings. But it is a fact that both species normally spend their time away from the breeding grounds continuously on the wing. Both have feathers on their tarsi (lower leg bones) to keep their legs warm at night. Swifts flying over the Alps at night, observed from gliders, drift in circular movements, beating their wings for a few seconds then gliding. On radar they can be traced rising gradually in the

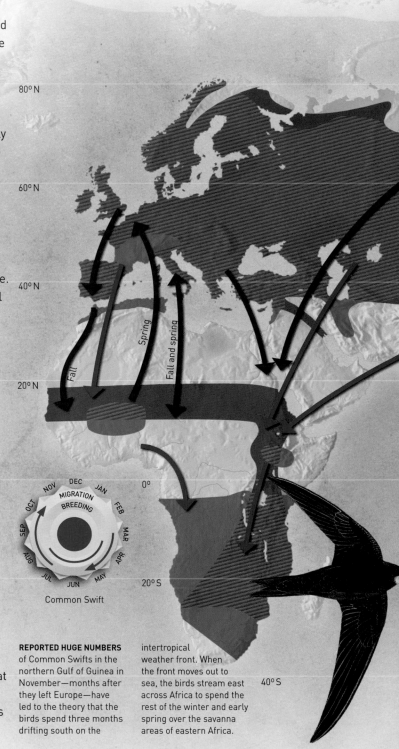

Common Swift

REPORTED HUGE NUMBERS of Common Swifts in the northern Gulf of Guinea in November—months after they left Europe—have led to the theory that the birds spend three months drifting south on the intertropical weather front. When the front moves out to sea, the birds stream east across Africa to spend the rest of the winter and early spring over the savanna areas of eastern Africa.

	**COMMON SWIFT** *Apus apus* **Wingspan** 18 in / 46 cm **Weight** 1⅓–1½ oz / 40–45 g **Journey length** 1,850–7,500 miles / 3,000–12,000 km	
	**SAND MARTIN** *Riparia riparia* **Wingspan** 11 in / 28 cm **Weight** ⅖ oz / 12.5 g **Journey length** 950–6,000 miles / 1,500–10,000 km	
	**BARN SWALLOW** *Hirundo rustica* **Wingspan** 13 in / 32 cm **Weight** ⅔ oz / 19 g **Journey length** 0–7,500 miles / 0–12,000 km	

Barn Swallow

Sand Martin

THE RETURN OF SWALLOWS to Europe in April marks the successful completion of a round-trip some 12,500 miles (20,000 km) long. If feeding trips are included, from the time they left in August or early September, the birds may have flown up to 190,000 miles (300,000 km).

Successful breeding birds return to the same area, perhaps even the same nest; young birds, also, tend to return to the area of their birth.

evening sky over towns and cities, and may drift with the wind until they have moved 20–30 miles (30–50 km).

Although Barn Swallows have been known migrants for generations, they are so widespread that it was only with the advent of banding that their winter homes could be pinpointed. Birds from central Russia join British breeders in South Africa; those from Germany winter in Zaïre; other areas are favored by birds from elsewhere in Europe. Migrant Barn Swallows winter in areas where there may be more than a dozen species of local, resident swallows and martins, and confusion between them has led to regular reports of European Barn Swallows breeding in Africa. This has not been proved, although House Martins have occasionally stayed in Africa to breed.

For all these species, winter is the time of a slow molt, when the flight feathers are gradually renewed without impairing the birds' powers of flight. Like life on migration, life in winter is not always easy, although the birds are not attached to a nest and can wander around to find the best food and conditions. The most frequent problem in winter is unseasonal cold as they arrive in Africa.

IN RECENT DECADES Sand Martins have been vulnerable to winter drought, and populations throughout western Europe dropped dramatically by more than 75 percent during the winter of 1968/9. Although the species has partially recovered several times, there have been further falls, and numbers have never returned to the peak years of the mid-1960s.

SAND MARTINS breed in sandy banks (right). In developed areas, these natural sites are rare since the banks of streams and rivers are no longer left to erode unchecked. Quarries have now become regular nest sites, as have piles of sand already cleaned and ready for building use.

Wagtails and pipits

Yellow Wagtails shelter from the desert sun behind discarded oil drums.

The wagtails and pipits are a diverse family in terms of migratory strategy and include long- and short-distance migrants, those that simply move up and down mountains, and birds that are totally resident in one place.

The Gray Wagtail is a bird of mountain streams that often winters in the lowland areas around the Mediterranean. The White Wagtails of Scandinavia and other parts of northern Europe are migratory. Many of those in France and Britain, however, stay for the winter; others go to southern Europe or northern Africa. Some of the distinctive Icelandic White Wagtails winter in Britain; others simply pass through on their way south, some even reaching Senegal.

Yellow Wagtails are trans-Saharan migrants, traveling through western Europe and northern Africa to the Sahel—the region south of the Sahara desert that stretches eastward across the continent from Senegal. They are among the best-studied long-distance nocturnal migrants. As they move south through France and Spain using reedbed roosts, they gain weight, averaging an increase from their breeding weight of around $\frac{2}{3}$ ounce (17 g) to $\frac{9}{10}$ ounce (25 g) or more by the time they reach Spain. At this weight they could theoretically make a journey lasting about 60 hours—taking them across Iberia, northern Africa, and the Sahara. Whether they do this is still a matter of debate.

THERE ARE MANY records over the last 100 years of birds resting in shade during the day in the desert. Work in the 1980s and 1990s on such birds as Yellow Wagtails showed that they may be fit and fat and resting deliberately—not, as had been supposed, because they were exhausted, having made a navigational error or misjudged how far they had traveled.

Birds probably rest by day because temperatures are too high to allow them to metabolize water efficiently.

Birds that have made a mistake are thin and congregate at oases to feed during the day, rather than rest.

Yellow Wagtail

YELLOW WAGTAIL
Motacilla flava
Wingspan
10 in / 26 cm
Weight
²⁄₃–1 oz / 17–28 g
Journey length
600–5,000 miles /
1,000–8,000 km

WATER PIPIT
Anthus spinoletta
Wingspan
10 in / 26 cm
Weight
¾ oz / 24 g
Journey length
0–950 miles /
0–1,500 km

It may seem strange that birds should put on such comparatively huge quantities of fat before they migrate. Surely, the argument goes, it would be better to carry no surplus weight and be as lean and fit as possible. In fact, the extra weight is not a great burden for the birds, since calculating how far they will get on a full fat load is an instinctive skill (see pp. 20–21). It is a problem if they have to escape from predators, but this is rare: they would have to be flying close to the ground and encounter an Eleonora's Falcon (see pp. 92–93). A migrant at several thousand feet is safe.

There are several races of Yellow Wagtails, with males differing in head pattern and color over their range. The British race *flavissima* (yellowest) is the only one that is predominantly yellow, although birds of the central Russian race have yellow heads. Birds belonging to the most usual European race, known as blue-headed, have pale blue-gray heads and a white stripe above the eye. Other European birds have completely dark heads. Birds from all populations winter in Africa.

Among the pipits, Tree Pipits are long-distance migrants, leaving the woodland glades of Europe for the savannas of western Africa. Meadow Pipits are a common sight to birdwatchers and one of the migrant species most likely to be seen flying from northern Europe to the Mediterranean. Since they are coastal birds, many Rock Pipits, by contrast, are able to stay at home in winter, although birds breeding in Norway and Sweden must migrate to slightly warmer climates farther south. Many of the alpine Water Pipits are altitudinal migrants; others winter in southern Britain, often on watercress beds.

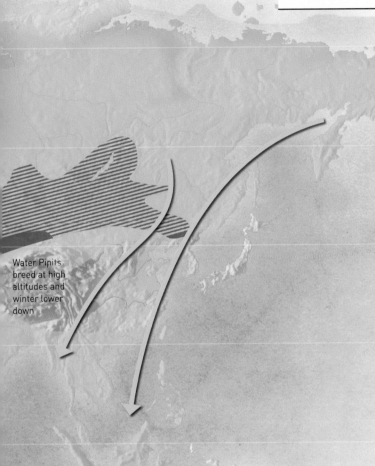

Water Pipits breed at high altitudes and winter lower down

Water Pipit

MANY WATER PIPITS spend their lives in a limited area close to the nest in which they were born. Like other species, they may find their breeding sites untenable in winter, with thick snow and ice from as early as August in some years, through to May or early June. A short flight away, and downward, however, are ideal feeding grounds for the whole winter. These areas can often support many birds, albeit in well-defended individual feeding territories. At night they may come together in large roosts in protected locations, such as stands of thick evergreen shrubs.

Such short-distance movements are not oriented in a single direction, and birds may go north or south. Long-distance flights are not oriented either. About 100 birds from the Alps winter in Britain each year. Presumably having survived one winter there, they return to familiar territory.

Chats and thrushes

These birds epitomize the advantages of migration over staying at home.

This group includes a wide variety of birds ranging in size from the European Robin, Redstart, and Nightingale to the large and bold Fieldfare (see pp. 128–29) and Mistle Thrush. Within Europe there are species that are long-distance migrants, birds that move relatively short distances within the continent, and many millions of birds that are resident. Often the residents are members of a species living in an area where they can survive year-round, while a different population of the same species, farther to the north and east, regularly moves great distances—forced to migrate by winter weather that would make survival at best improbable, at worst impossible.

The closely related Stonechat and Whinchat provide an interesting insight into the relative effects of long- and short-distance migration on the life history of birds. The Stonechats often remain in northern Europe for the winter—or move south only as far as western Europe or the Mediterranean basin. Whinchats, on the other hand, are trans-Saharan migrants, leaving Europe to spend the winter on the open savannas of Africa.

One immediate effect is that Stonechats are able to start breeding much earlier than Whinchats. Mid-March may see stonechats laying a first clutch, while their second coincides with

Common Nightingale

NIGHTINGALES have a reputation as specialist songsters of the night, but most also sing during the day. They are seldom heard by day since they have so many rivals then, but at night, without competition, their song is audible over great distances. These nocturnal habits also mean that many people familiar with nightingale song have never seen the bird.

Although many of their close relatives, including Bluethroats (see p.128), are brightly colored, Nightingales are warm rufous brown with a brighter tail. They are larger than European Robins, skulk in bushes and undergrowth, and do not share the Robin's upright stance.

Common Redstart

REDSTARTS nest in cavities or nest boxes in many areas; in others they breed in tunnels under vegetation. Nesting is carefully timed—they lay their bright blue eggs so that their young can take advantage of the high- summer glut of insects in deciduous woodland.

Male Redstarts are among the most handsome Eurasian small birds, with a dazzling white forehead and black on the face, chin, and upper breast.

80° N
60° N
40° N
20° N
0°

COMMON REDSTART
Phoenicurus phoenicurus
Wingspan
9 in / 23 cm
Weight
⅓–⅞ oz / 15–24 g
Journey length
1,300–3,750 miles /
2,000–6,000 km

COMMON NIGHTINGALE
Luscinia megarhynchos
Wingspan
10 in / 25 cm
Weight
⅔–1¼ oz / 19–36 g
Journey length
1,600–3,400 miles /
2,500–5,500 km

NORTHERN WHEATEAR
Oenanthe oenanthe
Wingspan
11½–12½ in / 29–32 cm
Weight
⅞–1½ oz / 24–41 g
Journey length
950–10,000 miles /
1,500–16,000 km

the first Whinchat brood. Stonechats have more third broods than Whinchats have second. This means that the potential for breeding, during a season, is much higher for a pair of Stonechats, and this is even further enhanced by the birds' choice of nest site. Stonechats often nest in gorse or other scrub; Whinchats, by contrast, choose to nest in grass and may have their nests in hay destroyed in the harvest.

Both species are in decline, but their current problems are due to loss of habitat and not to differences in lifestyle. Stonechats must have a high breeding potential, since in a cold winter, populations will suffer gravely. This ability to produce more young allows them to make up these losses over three or four years. Whinchats have an easier time, since they are efficient migrants and are able to maintain their populations at a steady level.

Long-distance migrants in this group include some birds, like the Redstart, that migrate mainly southwest through Europe, but others that migrate on a broad front. These include the Whinchat and Black-eared Wheatear. Eastern species, including the Thrush Nightingale and Isabelline Wheatear, migrate in an easterly direction around the Mediterranean. They are classic nocturnal migrants, and some seem to have specific weight-gaining areas. Most Redstarts stop off in Iberia for this purpose.

Robins, too, move by night, although many European populations are resident. Their autumn passage is sometimes spectacular, with huge numbers of birds appearing together in Belgium, the Netherlands, or Britain when the weather forces them to make an emergency landfall. Most winter in France, Spain, Portugal, and northern Africa. The return passage along the North Sea coast to the Scandinavian breeding grounds is usually concentrated in a short period in the first half of April.

EVOLUTION IN ACTION

At the height of the last glaciation, an ancestral wheatear was breeding farther south than now, in the same habitats used today, and nesting anywhere that was free from ice for long enough to rear a brood. As the ice retreated, possible breeding sites also moved north, east, and west, at a rate of some ½–1¼ miles (1–2 km) a year.

Since the original birds had migrated to sub-Saharan Africa, their descendants continued to return from their more distant breeding areas to their ancestral winter home. Had the ancestral species also wintered in Asia, both areas might have been used, or two species might have evolved.

THE STORY OF THE NORTHERN WHEATEAR is one of a supremely successful migrant that has expanded its breeding range across the whole of Europe and Asia. In the west, it has reached much of Greenland and coastal Canada; to the east it extends to Alaska and western Canada. This represents a circumpolar range of 320° and a north–south range from the Atlas Mountains at 30°N to Ellesmere Island at 80°N.

All these birds winter in sub-Saharan Africa, despite the fact that many could find suitable wintering areas far closer to their breeding grounds in Central America or Asia.

Northern Wheatear

Warblers

The long-term survival of many warbler populations is in doubt.

More warblers than birds of any other group leave Europe for Africa every fall. While swallows and martins dominate the wide open spaces of the air, warblers are birds of dense vegetation—sedge fields, reedbeds, scrubland, or woodland. Within this broad characterization, however, are individual warbler genera, each with its own habits and habitats.

Warblers of the *Acrocephalus* and *Locustella* genera, for example, frequent short, tangled, and often wet vegetation. The Grasshopper Warbler is a champion skulker, with a streaked coloration for easy concealment. Sedge Warblers are similar, but they are more adventurous in their habits. This species has been particularly badly affected by the winter weather in the Sahel: mid-1980s populations were around one-third of those of the mid-1960s.

The *Sylvia* warblers favor scrubland and, to a lesser extent, woodland in summer. The Common Whitethroat is equally at home in scrub and hedges. These long-tailed, relatively short-winged birds were once as numerous as Willow Warblers, but the entire Common Whitethroat population was reduced by 75 percent at a stroke in the winter of 1968/9.

Lesser Whitethroats, too, prefer scrub and hedges. All populations of this gray bird with a piratical black eyepatch winter in eastern Africa, the majority in Ethiopia, and all seem to follow the same route north in spring through the Near East and over the eastern Mediterranean. In the fall, however, British birds and perhaps others from western Europe take a more westerly route through Italy and across the Mediterranean into Egypt. Since they winter in a different area, Lesser Whitethroats have not declined in the same way as their Common Whitethroat cousins. Populations have fluctuated but remain within broad limits.

The two larger *Sylvia* warblers are similar in shape and size—but different in color. The Garden Warbler, a trans-Saharan migrant, is brown-olive with no distinguishing features. Blackcaps are brown-capped as juveniles, but when they molt, the males acquire glossy black caps. Southern European Blackcap populations are almost entirely resident, while Scandinavian breeders are long-distance migrants, wintering south of the Sahara. Birds from southern Germany winter in Britain and Ireland; those from Britain fly south to winter around the Mediterranean Sea.

Blackcaps are the only warblers whose populations have increased since the 1960s: numbers of birds frequenting woodland (their preferred habitat) have risen by 50 percent; those of birds on farmland by around one-third.

WARBLERS IN DECLINE

The major problem facing migrant warblers is the weather in the Sahel region, where several species spend the winter. Here disastrous droughts have severely affected the local plant, animal, and human ecology.

The Common Whitethroat winters in scrub and wooded savanna. After some years of drought, the scrub, which is resistant to arid conditions in the short term, was affected. Over the winter of 1968/9, conditions became critical, with the result that 75 percent of Common Whitethroats perished.

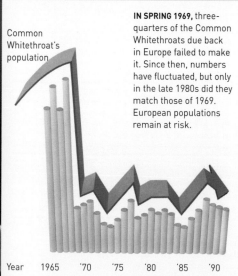

Common Whitethroat's population

IN SPRING 1969, three-quarters of the Common Whitethroats due back in Europe failed to make it. Since then, numbers have fluctuated, but only in the late 1980s did they match those of 1969. European populations remain at risk.

Year   1965   '70   '75   '80   '85   '90

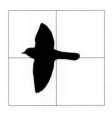

COMMON WHITETHROAT
Sylvia communis
Wingspan
8¼ in / 21 cm
Weight
½–¾ oz / 14–21 g
Journey length
1,250–5,600 miles /
2,000–9,000 km

REED WARBLER
Acrocephalus scirpaceus
Wingspan
8 in / 20 cm
Weight
⅓ oz / 11 g
Journey length
950–3,750 miles /
1,500–6,000 km

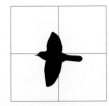

WILLOW WARBLER
Phylloscopus trochilus
Wingspan
8 in / 20 cm
Weight
³⁄₁₀–²⁄₅ oz / 8–12 g
Journey length
2,500–8,700 miles /
4,000–14,000 km

EASTERN REED WARBLER populations are grayer than the more rufous western birds and migrate in an easterly direction, skirting the Mediterranean.

When the migrants reach Europe in the spring, these birds are looking for marshy reedbeds in which to nest. The males join forces in rhythmic song at dawn to attract the later-returning females. It is to every male's advantage to have more females in the locality   the next female may be its mate for the year, may stop a neighbor from enticing a chosen mate away, or even provide a second hen for a bigamist.

Reed Warbler

Willow Warbler

ABOUT ONE-FIFTH of all migrants leaving Europe and Asia to winter south of the Sahara are Willow Warblers—almost a billion birds. This amazing number is, however, put into perspective by their small size—one billion warblers weigh the equivalent of, perhaps, 4,000 swans.

Their light weight enables these birds, and others of the *Phylloscopus* (leafpeckers) genus, to feed on the fringes of stands of vegetation and pick their food from the leaves.

Their sweet, descending song can be heard everywhere that is not totally open habitat or closed-canopy woodland.

Unusually, most Willow Warblers undergo one molt in late summer, before they leave the breeding grounds, and another in Africa in winter.

For reasons that are unclear, after 30 years of relative stability, the recorded number of Willow Warblers returning to Europe has declined each year since 1990.

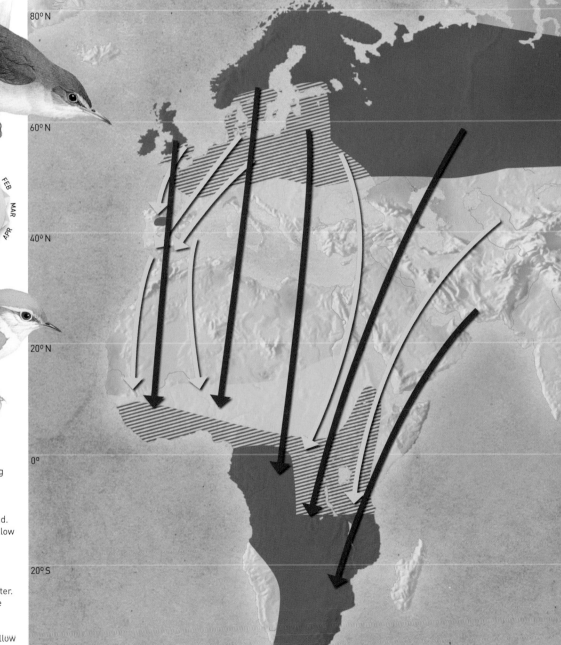

Shrikes

Conditions in the Great Rift Valley are ideal for migrant shrikes.

RED-BACKED SHRIKE
Lanius collurio
Wingspan
10¼ in / 26 cm
Weight
⅓ oz / 10 g
Journey length
2,500–6,800 miles /
4,000–11,000 km

The shrikes are predatory songbirds that feed on small birds, mammals, reptiles, and large insects. The colorful Red-backed Shrike is a southeasterly migrant, with all breeding populations heading over the eastern Mediterranean and down through the savanna areas of the Great Rift Valley. These birds winter in all but the southernmost parts of South Africa.

Today, the western limit of the breeding range of Red-backed Shrikes is in northwestern Spain; 100 years ago they were common summer residents in parts of Britain, but habitat loss and a general depletion in the numbers of large insects have pushed them south and east. They share this northwestern corner of Spain with the similarly sized and closely related Woodchat Shrikes, but elsewhere in Iberia only the latter breed.

Woodchat Shrikes are also long-distance migrants. Eastern races migrate to the south, and some may spend the winter in the extreme southwest of the Arabian peninsula. Western populations cross the Sahara to winter south of the desert in the Sahel. Their fall route takes them southwestward over the Iberian peninsula, but, like many other species, their return journey is more easterly—avoiding the generally wet weather on the fringe of the continent in late winter and early spring.

Every few years, when an anticyclone becomes established over southwestern Europe at the right time, prompting better than average spring conditions in northern Europe, Woodchat Shrikes overshoot their breeding grounds in southern France and Spain to reach northern France and Britain.

Red-backed Shrike

RED-BACKED SHRIKES are well adapted to the migratory way of life; since they are able to prey on their fellow migrants, they therefore move with them. These birds were once known as "butcher birds" because of their habit of storing surplus food—birds, mammals, and insects—on thorn bushes or barbed-wire fences.

Although these birds are common in France and Germany, reports of these birds in Britain are rare: this area is outside their usual range. Barred, brown young from the Scandinavian breeding populations may travel as far west as Britain on their fall explorations (see pp. 28–29).

Flycatchers

Cool, wet springs force birds east.

PIED FLYCATCHER
Ficedula hypoleuca
Wingspan
9 in / 23 cm
Weight
½–⅔ oz / 12.5–19 g
Journey length
1,250–4,350 miles /
2,000–7,000 km

SPOTTED FLYCATCHER
Muscicapa striata
Wingspan
9½ in / 24.5 cm
Weight
½–1 oz / 15–28 g
Journey length
2,200–8,000 miles /
3,500–13,000 km

There are only two species of flycatcher widely distributed in western Europe: the Spotted, a bird of open woodland and gardens, and the Pied, a truly woodland species. Both have broad bills with well-developed bristles around them that literally extend the size of the mouth, making it easier for them to catch insects in flight. These bristles look like stiff hairs, but they are in fact modified feathers. Both species forage by sitting on a perch and keeping watch for flying insects—a quick dash and grab and back to the perch. This strategy presumes that there are plenty of large flying insects, which cannot be guaranteed until summer is well advanced.

In fact, Pied Flycatchers can also feed successfully on the ground, rather like Robins. This means that they can return to the oak woods of northern Europe as early as mid-April, at least three weeks in advance of the Spotted Flycatchers. This is essential for successful breeding and is timed to coincide with the superabundant caterpillar crop of the woodland spring. Pied Flycatchers breed in holes, for which they are in direct competition with resident tits. This may keep the former away from lowland areas in some parts of Europe and confined to higher ground, where tits are not easily able to survive the winter.

Migrating Pied Flycatchers have been studied in detail at their staging post in northwestern Portugal. Each bird has its own feeding territory, which it defends assiduously. Migrants unable to find a territory are excluded from the best feeding areas until the initial holder has put on enough weight to allow it to migrate southward.

These birds have a carefully arranged circuit of their home range because the large insects on which they feed may lie motionless for 15–20 minutes if they detect the presence of a flycatcher. The territory owner comes back every half hour or so to take the insects that are, by this time, flying around again. A flycatcher interloper would disturb this precisely timed routine.

Spotted Flycatchers are long-distance migrants, with birds from Finland and Wales discovered wintering in South Africa and an Irish breeder spending the winter in Angola. The species is one that shows a migratory divide at about 12°E, with birds breeding to the west migrating westward through Spain and those to the east traveling through the eastern Mediterranean. Although this means that birds from western Europe almost all travel to western Africa, once they have crossed the Sahara, they may continue to move east and south. To date, 35 birds banded in Europe have been found in the Zaïre River basin.

♀

♂

Pied Flycatcher

NOV DEC JAN
OCT MIGRATION FEB
BREEDING
SEP MAR
AUG APR
JUL JUN MAY

Spotted Flycatcher

THE DAPPLED LIGHT AND SHADE of summer woodland tends to make male Pied Flycatchers appear pure black and white to observers. The birds do, however, have a good deal of brown coloration, although males are darker than females.

Most Pied Flycatchers pass through a staging post in northern Portugal in the fall where, for a couple of months, they may be the most common birds of the cork oak forests. The return route in spring is different, with birds taking a more easterly course. At this time of year, the climate in northern Portugal is cold and wet, and completely unsuitable for a flycatcher.

THE MOLT OF SPOTTED FLYCATCHERS is unusual in that, unlike almost all other perching birds, the first of the primary feathers (the main flight feathers) to be replaced is the outermost and not the innermost. No logical reason has been found for this, but the Spotted Flycatcher is a species that relies on its powers of flight for feeding. Unlike the swallows and martins, these birds tend to hug cover, and therefore their wing feathers are susceptible to wear.

The molt is undertaken on the wintering grounds of Africa, Arabia, or northwestern India in the middle of the season, usually between November and February.

Winter visitors from the far north

The short Arctic summer, lasting no more than three to four months, offers a brief window of opportunity for birds to arrive, breed, and then depart. There are risks involved in breeding in the far north: the fickle weather, which can produce snow in July, the variability of the food supply, and, not least, the need to migrate long distances to winter in warmer climes. But for those birds prepared to overcome these difficulties, the Arctic offers extensive areas of suitable habitat, a relatively low number of predators compared with regions farther south, and, despite fluctuations, an abundant food supply that, very importantly, is available 24 hours a day in this land where the summer sun never sets.

Most of the birds nesting in the Arctic are large—predominantly waterfowl and shorebirds—and adapted for the necessary long migrations. Their populations can also withstand the occasional breeding failure that arises when severe weather sets in, even in the middle of the summer, killing off the superabundance of insect life that flourishes and often covering the vegetation with snow.

Snow Geese nest in the low-lying, boggy Arctic tundra, exploiting the ideal nesting and feeding habitat it affords, but they must fly south for the winter.

Patterns of migration

Birds that breed during the brief Arctic summer must head south for the winter.

The need to escape the Arctic winter forces almost all the birds breeding in that region to leave—some as early as August, the rest throughout September and October—and make for somewhere warmer with an assured food supply. The obvious direction is south, but that is, in fact, only a general course. Many birds more specifically move southwest or southeast to reach the wintering area of their choice. In some cases this direction can reveal how a particular species may have colonized the Arctic.

Many birds breeding in the Arctic make the shortest possible journey to the nearest suitable habitat each fall. Some Snowy Owls, for example, winter within the southern parts of their breeding range or move only a few hundred miles to the south, while Purple Sandpipers stay as far north as they can, in Iceland and northern Norway, and move no farther than the nearest ice-free coast.

Other Arctic breeders, however, migrate thousands of miles, several crossing the equator and reaching the southern shores of South America, South Africa, and Australia. It is difficult to be certain of the driving factor in every case of long-distance migration. Certainly the need to find the right habitat with an adequate food supply may cause birds to overfly potentially suitable areas that already hold large numbers of species that might provide unacceptable competition.

Some journeys are almost certainly governed by historical factors. Instead of taking what would seem to be the obvious route south through North America, several birds breeding in Greenland and on the numerous islands of the Canadian Arctic

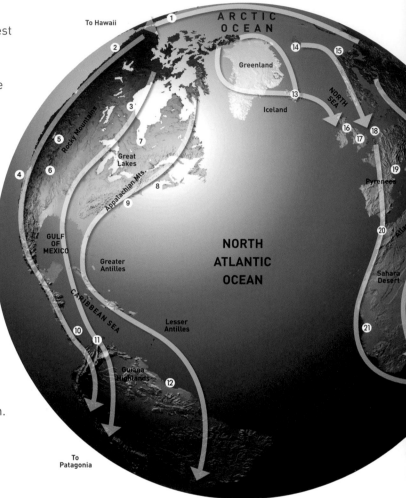

✪ HOT SPOTS

Arctic breeders are often long-distance migrants; as a result every continent affords at least one prime site where these birds can be seen.

1. St. Paul Island, Alaska
2. Copper River delta, Alaska
3. Churchill, Manitoba
4. San Francisco Bay, California
5. Salt Lake, Utah
6. Cheyenne Bottoms, Kansas
7. James Bay, Ontario
8. Bay of Fundy, New Brunswick/Nova Scotia, Shepody Bay, New Brunswick
9. Delaware Bay, Delaware
10. Panama
11. Los Olivitos, Venezuela
12. Coast of Surinam
13. Lake Myvatn, Iceland
14. Ny Alesund, Svalbard
15. Varangerfjord, Norway/Lake Inari, Finland
16. Solway Firth, Scotland
17. The Wash, England
18. Waddenzee, The Netherlands
19. Camargue, France
20. Coast of Morocco
21. Banc d'Arguin, Mauritania
22. Danube delta, Romania
23. Volga delta, Russia
24. Kyzylagach NR, Azerbaijan
25. Bharatpur, India
26. Kamchatka peninsula, Russia
27. Mai Po marshes, Hong Kong

ABOUT 100 BIRD SPECIES breed on the Arctic tundra in the brief northern summer, and all but perhaps half a dozen exceptionally hardy species migrate. About 30 of these species are waterfowl, another 30 or so shorebirds.

THE WANDERING TATTLER'S migration takes it from the mountain streams of Alaska and eastern Siberia to the Pacific coasts of South America, Australia, New Zealand, or Japan—a journey of up to 9,300 miles (15,000 km).

THE WORLD'S 2,000–2,600 Spoon-billed Sandpipers reed in northeastern Siberia and Kamchatka and migrate along the Pacific coast of Asia to winter in Southeast Asia, from eastern India to southern China.

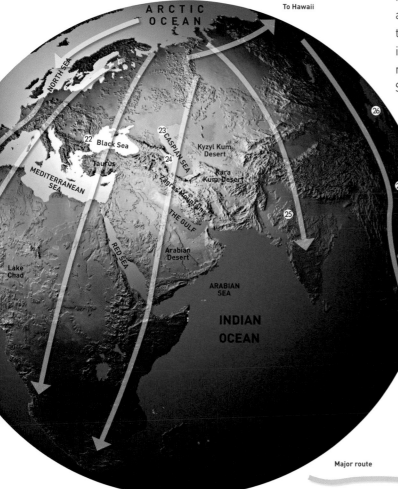

head southeast to western Europe. At the end of the last ice age, 10,000 years ago, the Red Knot, Northern Wheatear, and the light-bellied race of the Brant, among others, were breeding in what is now western Europe. As the ice retreated, most birds nesting in this area moved steadily north and northeast to colonize Scandinavia and Russia. But a few species successfully island-hopped northwestward, via the Faroes and Iceland to Greenland, and then on to Canada. Throughout the thousands of years of this gradual expansion, the birds maintained their traditional pattern of migration to the southeast. And, although the distances involved may seem great on a map—and such a journey involves overflying or skirting the Greenland ice cap—on a globe they do not seem so daunting.

The majority of Arctic-breeding species are waterfowl and shorebirds, and their need for watery habitats, both as migration stopovers and as winter homes, dictates the directions they take and destinations they target. Many of these birds are channeled to the coasts of North America and Eurasia, or follow one of two major flyways: via the Great Lakes and down the Mississippi valley, or from the White Sea to the Baltic and into the North Sea before heading down the coasts of France and Iberia toward western Africa.

THE BREEDING GROUNDS of the far north fringe the Arctic Ocean and extend south roughly to the Arctic Circle, although in parts of Canada and Russia areas that can be described as Arctic exist well to the south of this line. Many birds fly almost due south over the continental landmasses, while the coastlines of North America and Eurasia provide easy-to-follow routes that are also approximately north–south in orientation.

A bird's route from the Arctic is primarily determined by what it faces on the first stage of its journey. Can it move slowly and steadily south, feeding as it goes, or is it faced with a long flight over unsuitable terrain or the sea where feeding opportunities are severely restricted?

Many birds breeding in the Canadian and Russian Arctic can move south across the continental landmasses, making comparatively short flights between resting places. But not all the land is favorable for birds, particularly in the eastern half of Russia, where the Gobi and other extensive deserts and high mountain ranges such as the Himalayas straddle the routes. Birds must cross such areas nonstop.

Species nesting in the far east of Siberia or in Alaska have the choice of a long journey over land, in which case they must head almost due west or east, respectively, before they can turn to the south, or they must commit themselves to a flight over the sea before making a landfall. The distance involved—from Alaska to California, for example—may be up to one-third shorter but there is no food to be found on the way.

Birds nesting in Greenland, Iceland, and Svalbard have no such choice—the first leg of their migration has to involve an over-sea flight of hundreds of miles.

Swans

Keeping the family together brings rewards on the wintering grounds.

One species of small swans and two species of large ones breed in the far north. The small Tundra Swan occurs widely in North America, where there is a race called the Whistling Swan, and across Europe and Asia, where the race is known as the Bewick's Swan. The two larger species, up to 70 percent bigger than the Tundra Swan, are the Trumpeter Swan, which has a restricted range in Alaska, northwestern Canada and the U.S., and the Whooper Swan, which breeds from Iceland through northern Europe and Asia to eastern Siberia.

All these swans travel long distances from their tundra breeding grounds to the coastal and fresh waters of temperate latitudes to the south. Many of their winter haunts are traditional and have been used for almost as long as reliable records exist. Moreover, the same swans return year after year to the same site and use the same stopping places on migration.

This habituation makes sense for the birds: if they are familiar with an area, it is easy for them to find food and to learn where

danger threatens. The pattern of returning year after year is passed on from one generation to the next. Since young swans take time to reach maturity, both in terms of size and in their ability to look after themselves, they are still dependent on their parents at the onset of their first Arctic winter. As a result, and contrary to the practice adopted by many species, the family migrates together so that the young learn the route, the places to stop on the way, and the ultimate wintering site. The following fall, after they have separated from their parents, the juvenile swans are able to follow the same route to the same destination and so the tradition continues.

It is not unusual for the young of the previous year to meet up with and become reattached to their parents, even though the adult birds may have a new brood of young with them. A "super-family" then develops, which may even contain the young of two or three previous years. There are definite advantages in belonging to a super-family: larger families reign supreme, dominating both

TUNDRA SWAN	TRUMPETER SWAN
Cygnus columbianus	*Cygnus buccinator*
Wingspan	**Wingspan**
6 ft–6 ft 11 in / 1.8–2.1 m	7 ft 3 in–8 ft 2 in / 2.2–2.5 m
Weight 7 lb 4 oz–16 lb 8 oz /	**Weight**
3.3–7.5 kg	15 lb 7 oz–27 lb 9 oz /
Journey length	7–12.5 kg
1,600–3,000 miles /	**Journey length**
2,500–5,000 km	0–1,600 miles / 0–2,500 km

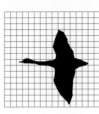

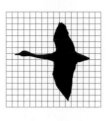

Tundra Swan

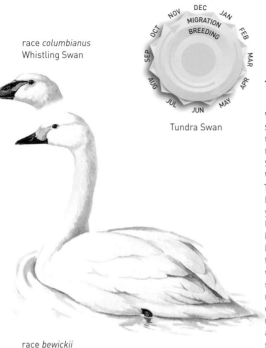

race *columbianus*
Whistling Swan

race *bewickii*
Bewick's Swan

WHISTLING AND BEWICK'S SWANS are subspecies of the Tundra Swan. A long-term study of Bewick's Swans by the British Wildfowl and Wetlands Trust identified individuals by their unique black and yellow bill patterns and learned much about the birds' interrelationships, behavior, and ability to return to the same wintering site. Recent studies using satellite radios have produced detailed information on migration routes, timing, and the use of stopover sites on the way.

smaller ones and pairs without any young, and keeping them away from the best feeding places.

Whooper and Trumpeter swans, which can measure up to 6 feet (1.8 m) from beak to tail, are around the maximum size for long-distance movement. Large birds cannot put on as much fat to use on migration as smaller ones, since they are already close to the heaviest weight their wings can lift off the ground (see pp. 24–25). A nonstop over-sea flight from Iceland to Britain of 550–750 miles (900–1,200 km) is approaching the limit for Whooper Swans, which often migrate in a strong tail wind to reduce the time they must spend in the air. These birds have also been recorded migrating at high altitudes, on occasion reaching 28,000 feet (8,500 m), where the winds are usually strongest.

THE TRUMPETER SWAN (left), hunted throughout the 19th century for its meat, skin, and feathers, was on the verge of extinction earlier this century. Conservation began in 1918 when hunting was outlawed. More recent measures include establishing birds raised in captivity in former breeding haunts and feeding flocks in winter. There has clearly been a good recovery, from some 2,000 birds in the early 1960s to nearly 25,000 birds today.

Geese ①

Barnacle Geese find lush winter pastures—but at the expense of farmers.

There are three populations of Barnacle Geese, each with discrete breeding and wintering ranges. The largest, over 350,000 birds, breeds mainly in Arctic Russia and winters in the Netherlands, Denmark and Germany. The smallest, about 28,000, breeds in Svalbard and winters on the Solway Firth, Scotland. The third population, about 60,000, breeds in east Greenland, wintering on islands off Scotland and Ireland, scattered between Orkney and County Kerry.

The separate populations use traditional migration stopovers. These sites are always in areas where there is an abundance of good grass so that the birds can put on sufficient fat reserves to complete the fall and spring migrations and, in spring above all, to cope with the physical stresses of the nesting period.

Many of the birds wintering in the Netherlands stop on Gotland, off the east coast of Sweden, in spring. During the 1960s, small numbers began to stay there through the summer, eventually nesting. This habit grew rapidly during the 1970s and 1980s, and today more than 1,500 pairs of Barnacle Geese nest on Gotland

and join the Russian birds as they pass through in the fall en route to the Netherlands.

Those Barnacle Geese that winter on the Solway Firth stop over in spring on the hundreds of small grassy offshore islands of northern Norway. They spend about three weeks in May feeding intensively in the long hours of daylight, before crossing the Barents Sea to Svalbard. The fall stopover is different: Bear Island, midway between Svalbard and northern Norway. From there, they undertake a single flight of more than 1,250 miles (2,000 km) to reach the Solway Firth.

Iceland is an ideal stopover for the Greenland-nesting geese, the lush pastures of its northern valleys providing good feeding in both spring and the fall. The winter distribution of these birds is unusual in that one location, the island of Islay in the Inner Hebrides, supports 60 percent of the total population; the remainder is scattered over a further 50 or 60 haunts. Numbers on Islay can reach 40,000 birds; the next largest flock is some 2,500, on the Inishkea Islands, Ireland. Many haunts hold only

BARNACLE GEESE, migrating over the sea, fly nonstop. When they finally make landfall, they first bathe and drink, then preen briefly before tucking their heads under their wings and going to sleep. Those that fly from Iceland to Islay in the fall, a journey lasting perhaps 20 hours, sleep for three or four hours before flying to the nearest pastures to feed.

		BARNACLE GOOSE			**BRANT**
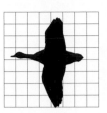		*Branta leucopsis*	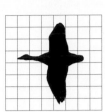		*Branta bernicla*
		Wingspan 4 ft 4 in–4 ft 9 in / 1.32–1.45 m			**Wingspan** 3 ft 7 in–3 ft 11 in / 1.1–1.2 m
		Weight 3 lb 11 oz–4 lb 13 oz / 1.4–2.2 kg			**Weight** 2 lb 10 oz–3 lb 5 oz / 1.2–1.5 kg
		Journey length 1,100–2,000 miles / 1,800–3,200 km			**Journey length** 1,900–4,000 miles / 3,000–6,500 km

Brant

PARTS OF THE BREEDING RANGE of the Brant are farther north than those of any other goose—on the Arctic islands of Canada and Russia. Many populations fly more than 2,500 miles (4,000 km) to reach their winter homes. Birds nesting on Wrangel Island in northeastern Siberia fly 4,000 miles (6,500 km) to western Mexico and may cover some 3,000 miles (4,800 km) of that in a single over-sea flight from Alaska.

100 to 200 birds. This uneven distribution requires careful management, not only because whatever happens to the birds on Islay dramatically affects the overall population, but also because this concentration in one place can create problems.

Barnacle Geese feed on the best grass they can find, which on Islay is the same grass that the farmers are growing for their sheep and cattle. After many years of conflict, when shooting geese was the only way of reducing the damage they were doing to the grass, a scheme was introduced in October 1992 whereby the farmers are paid by the government conservation agency to tolerate the geese, whose numbers, though high on Islay, are internationally vulnerable.

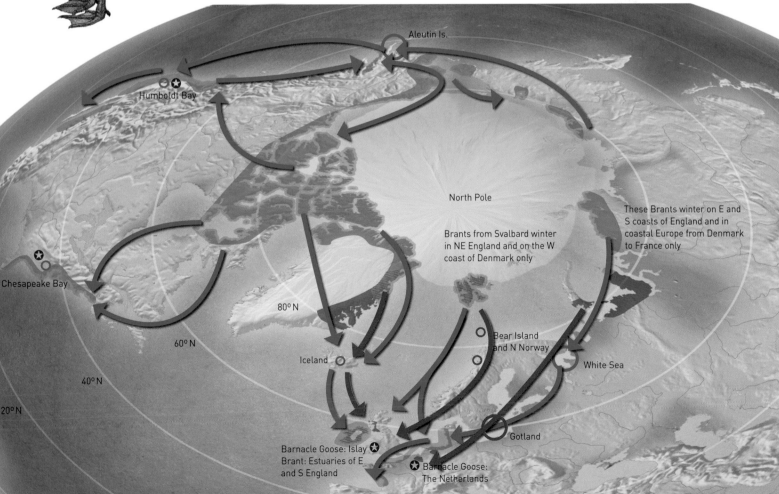

Aleutin Is.

Humboldt Bay

North Pole

These Brants winter on E and S coasts of England and in coastal Europe from Denmark to France only

Brants from Svalbard winter in NE England and on the W coast of Denmark only

Chesapeake Bay

80° N

60° N

40° N

20° N

Iceland

Bear Island and N Norway

White Sea

Gotland

Barnacle Goose: Islay
Brant: Estuaries of E and S England

Barnacle Goose: The Netherlands

Geese ②

More than eight million Snow Geese leave the far north for the winter.

Arctic Canada is the breeding ground for between six and eight million Lesser Snow Geese, about 800,000 Greater Snow Geese, and one million of the smaller Ross's Snow Goose. There are also 85,000 Lesser Snow Geese in eastern Siberia. All migrate in winter.

Snow Geese breed in colonies of between 1,000 and 150,000 pairs. Their white plumage, relieved only by black wing tips, means that they cannot rely on camouflage to avoid detection by predators when nesting on the open tundra. Instead, they depend on the sheer number of individuals in the colony to intimidate would-be attackers. Only birds at the edges of the group are at risk; these are likely to be the immature or less fit members of the colony.

A feeding flock of geese might appear to be without structure, but the family parties and pairs stay together, moving through the

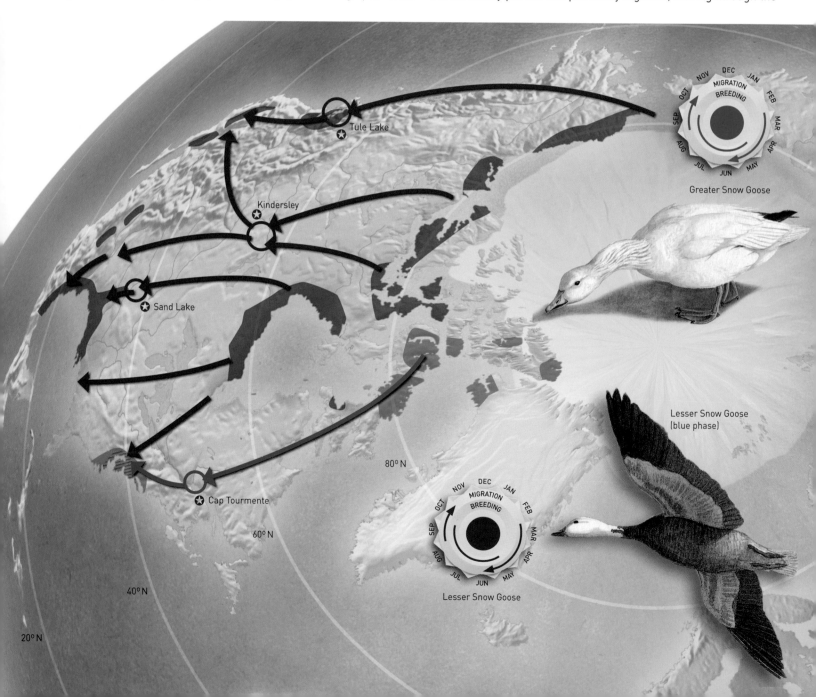

Greater Snow Goose

Lesser Snow Goose
(blue phase)

Lesser Snow Goose

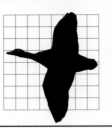

LESSER SNOW GOOSE
*Anser caerulescens
caerulescens*
Wingspan
4 ft 4 in–5 ft 3 in / 1.32–1.6 m
Weight
4 lb 7 oz–6 lb 10 oz / 2–3 kg
Journey length 1,250–
3,000 miles / 2,000–5,000 km

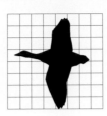

GREATER SNOW GOOSE
*Anser caerulescens
atlanticus*
Wingspan 4 ft 9 in–5 ft 9 in /
1.45–1.75 m
Weight 6 lb 3 oz–9 lb 4 oz /
2.8–4.2 kg
Journey length 2,200–3,000
miles / 3,500–5,000 km

flock as distinct units. Each group keeps an area around it clear of other geese so that competition for food is reduced. The adult male of a family is especially active in making sure that other geese do not come too close.

All geese pair for life during their second winter, when they are between one and two years old and still immature. The males display to the females, each of which selects a mate larger than she is. It is an obvious advantage to a female to have a mate able to defend her and the nest against both predators and the interference of other geese.

Color, too, plays a part in the process of selecting a mate. Lesser Snow Geese occur in two distinct, genetically controlled color types, or phases—white and blue (in fact, more a gray color). The female's choice is thought to be influenced by the color of her parents.

THE IMPACT OF EVEN A SINGLE GOOSE on a jet airliner can be disastrous. For this reason, observers south of Canada's Winnipeg airport monitor the northward progress of Lesser Snow Geese and warn air-traffic controllers of their approach. If necessary, the airport can be closed down for hours, or occasionally days, during the peak of migration.

Winnipeg lies on one of North America's major migration routes for northern breeders. The migratory patterns of Snow Geese and many other species are influenced by the continent's geography (see also pp. 52–53). In general, birds breeding in the eastern Canadian Arctic migrate to the Atlantic seaboard; those breeding in central Canada use the Mississippi flyway to reach the Gulf of Mexico; the more westerly breeders, particularly those in Alaska, migrate along, and winter close to, the Pacific coast.

Lesser Snow Geese breeding around Hudson Bay follow this pattern, migrating south across the Great Lakes and down the Mississippi valley to winter on the extensive coastal marshlands and low-lying farmland bordering the Gulf of Mexico. In winter, and particularly as spring approaches, the birds must obtain enough food to put on fat both for the return migration and for at least the first week or two back at the breeding grounds, when it is often still too cold for vegetation to grow.

Shorebirds ①

Estuary stopovers are vital for most of these long-distance migrants.

There are about 215 different species of shorebirds spread throughout the world, more than half of which migrate. While some of their journeys are comparatively short—the Wrybill, for example, breeds on South Island, New Zealand, and winters on North Island— there are at least 65 species that breed in the northern hemisphere and winter wholly or partly within the southern hemisphere.

The Arctic tundras of Europe, Asia, and Canada offer a safe breeding habitat for the vast majority of shorebirds that cross the equator from their winter homes in Africa, South America, and Australasia. These include the Curlew Sandpiper, which breeds in northern Siberia on the shores of the Arctic Ocean and winters in southern Africa, India, and Australia, and the White-rumped Sandpiper, which nests on islands in the Canadian Arctic and migrates some 8,700 miles (14,000 km) to winter in Chile and Argentina.

Several species, including the Sanderling and Black-bellied Plover, breed right around the North Pole in the North American and Eurasian Arctic. Both are equally far-flung in winter, when there are populations wintering on the coasts of South America, South Africa, and Australasia.

Although shorebirds are associated with water, feeding on tidal mud flats and in shallow fresh waters, most do not swim, making them as vulnerable on sea crossings as land birds. That they are able to cope with long ocean journeys stems from their flying ability, which is greater than that of many small land birds. The long, narrow wings of most migrant shorebirds enable them to fly quickly and efficiently, and so travel farther than a less efficient bird of equal size for each calorie of energy used.

Some species fly directly, over land and sea, between their breeding and wintering grounds; others must stop to refuel on the way. Shorebirds cannot feed at sea and have to make a landfall to find food. For many species, migration routes are dictated by the location of estuaries and coastal wetlands, where they find the invertebrates on which they feed.

The importance of these areas to migrating shorebirds is becoming increasingly clear. An estuary staging post is as vital to a migrant for the few days to few weeks that it spends refueling there as it is to a bird that stays for the whole winter. If a migrant cannot find enough food to enable it to complete the next leg of its journey, it will not survive.

Hawaii

Spring

Fall

S limit of breeding range

VAST FLOCKS of Eskimo Curlews once migrated down the Atlantic coast of North America, en route between the Canadian Arctic and the pampas of South America. They were shot in such numbers as to be "scarce" in the 1890s, and by 1930 the Eskimo Curlew was thought— mistakenly—to be extinct. A small number survives today.

NOV DEC JAN
OCT MIGRATION FEB
     BREEDING
SEP   MAR
AUG   APR
   JUL JUN MAY

Eskimo Curlew

ESKIMO CURLEW
Numenius borealis
Wingspan
32–33 in / 81–85 cm
Weight
7¾–10½ oz / 220–300 g
Journey length
6,000–8,700 miles /
10,000–14,000 km

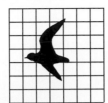

PACIFIC GOLDEN PLOVER
Pluvialis fulva
Wingspan
24–26 in / 60–67 cm
Weight
3¼–5¼ oz / 90–150 g
Journey length
3,000–8,000 miles /
5,000–13,000 km

RUFF
Philomachus pugnax
Wingspan
18–23 in / 46–58 cm
Weight
1¾–2¾ oz / 50–80 g
Journey length
2,500–9,300 miles /
4,000–15,000 km

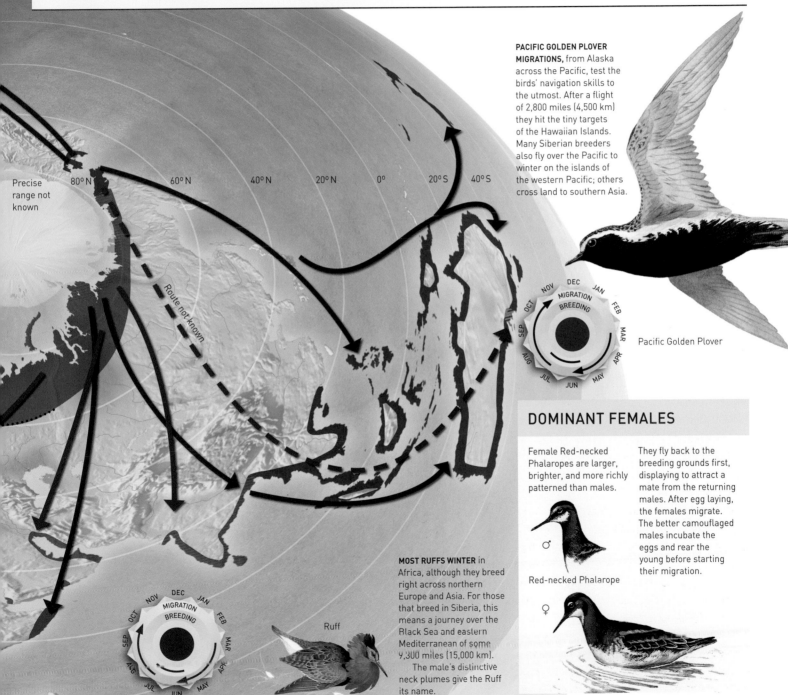

PACIFIC GOLDEN PLOVER MIGRATIONS, from Alaska across the Pacific, test the birds' navigation skills to the utmost. After a flight of 2,800 miles (4,500 km) they hit the tiny targets of the Hawaiian Islands. Many Siberian breeders also fly over the Pacific to winter on the islands of the western Pacific; others cross land to southern Asia.

Precise range not known

Route not known

Pacific Golden Plover

DOMINANT FEMALES

Female Red-necked Phalaropes are larger, brighter, and more richly patterned than males.

They fly back to the breeding grounds first, displaying to attract a mate from the returning males. After egg laying, the females migrate. The better camouflaged males incubate the eggs and rear the young before starting their migration.

Red-necked Phalarope

MOST RUFFS WINTER in Africa, although they breed right across northern Europe and Asia. For those that breed in Siberia, this means a journey over the Black Sea and eastern Mediterranean of some 9,300 miles (15,000 km).

The male's distinctive neck plumes give the Ruff its name.

Ruff

Shorebirds ②

Widely traveled Red Knots are on the move for up to seven months each year.

Routes across or around S America not known

Delaware Bay

Bay of Fundy

Surinam coast

Iceland

Banc d'Arguin

Waddenzee and estuaries of N France and S and E England

Red Knots are among the most northerly breeding shorebirds, nesting on the Arctic islands of Canada and Siberia closest to the North Pole, and on the extreme north coast of Greenland. This far north the summer is short and the birds may be on the breeding grounds for only two months: June and July. Although they add plant shoots to their usual diet of invertebrates at this time, there is not enough food to allow them to arrive earlier or stay longer.

The advantages of breeding so far north are the lack of competition from other species for nest sites and food, and the low density of predators. Red Knots further reduce the risks of competition and predation by nesting scattered over the tundra, with families rarely less than ⅝ mile (1 km) apart. This behavior changes on migration and in winter, however, when they form flocks tens of thousands strong.

In addition to nesting farther north than other shorebirds, many Red Knots winter far south of the equator, reaching the southern tips of the Americas and Africa and the coasts of Australia and New Zealand. But not all journey so far; many stay north of the equator, congregating on the coasts of northwestern Europe and around the Gulf of Mexico.

Many Red Knots do not take the most obvious route from their breeding grounds, but head in winter for sites that are, perhaps, more distant than those used by birds from other nesting areas. Eastern Canadian breeders, for example, do not head south to the Gulf coast of the United States—the northern limit of the wintering range of central Canadian birds, and obviously suitable winter habitat. Instead, they fly southeastward, across Greenland and the North Atlantic to wintering areas in northwestern Europe.

This and other "longer-than-necessary" journeys are usually the result of birds spreading and colonizing new areas while retaining traditional migration routes. Eastern Canada, for example, was colonized by birds from Greenland that had established this southeasterly migration route.

There is a similar migratory divide in eastern Siberia. Birds breeding in the far east fly south to Australia, while those breeding in central Siberia head southwest to western Europe before journeying on to western and southern Africa.

Red Knots frequently journey up to 1,900 miles (3,000 km) nonstop, often across open sea. Thus it is vital for them to put on enough fat for fuel. Birds leaving Britain in spring increase their weight by 50 percent before crossing to Iceland, and lose most of that en route. They feed again in Iceland before the flight on to Canada.

WATCHING THE SHOREBIRDS

Red Knots, in common with other shorebirds, winter on estuaries and in coastal lagoons where small mollusks and crustaceans—the main elements of their diet at this time of year—abound. The largest winter concentration of these birds is on the Waddenzee in the Netherlands, followed by the Wash in Britain. Most British estuaries—the Thames, Severn, Dee, Solway Firth, and Morecambe Bay—are shorebird sites.

At peak passage times, almost any estuary or stretch of inland mud may attract migrant shorebirds to rest and feed, though the sites listed here are especially important. In the fall, up to 2.5 million shorebirds pass through the Bay of Fundy, Canada, while some 2 million winter on the Banc d'Arguin, Mauritania.

High tide is the best time to watch these birds, when incoming water forces them inshore. They become concentrated together in spectacular flocks—100,000 birds may crowd on to a few hundred yards of shoreline.

80° N   60° N   40° N   20° N   0°   20° S   40° S

North Pole

Yellow Sea coast ✪

Port Philip ✪
Bay / Corner Inlet

✪ Eighty Mile
Beach /
Roebuck Bay

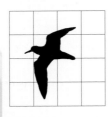

RED KNOT
Calidris canutus
Wingspan
22–24 in / 56–61 cm
Weight
3⅓–7½ oz / 95–215 g
Journey length
1,500–10,000 miles /
2,500–16,000 km

IN THE COURSE OF A YEAR, Red Knots spend time in several locations in addition to their breeding and wintering areas. Birds that breed in the eastern Canadian Arctic leave in July and fly to Iceland, where they spend August. By September they are in southern Norway, where they stay for a few weeks before flying across the North Sea to winter on the Wash in England or the Waddenzee. They head north again in March, stopping in Iceland during April and in western Greenland in May, arriving back on their Arctic breeding grounds in early June.

Red Knot

OTHER PRIME SITES INCLUDE
Delaware Bay, New Jersey, U.S.
Copper River delta, Alaska, U.S.
Gray's Harbor, Washington, U.S.
Cheyenne Bottoms, Kansas, U.S.
Camargue, France
Baie l'Aiguillon, France
Coto Doñana, Spain
Faro, Portugal
Coast of Morocco
Walvis Bay, Namibia
Mai Po marshes, Hong Kong
Gulf of Thailand
Shinhama Preserve, Tokyo, Japan
Eighty Mile Beach and Roebuck Bay, Western Australia
Gulf of Carpentaria, Northern Territory, Australia

Thrushes

Migrants blown off course have founded a new, sedentary colony.

Wheatears, redstarts, chats, the European Robin, and the Bluethroat, as well as the larger "true" thrushes, all belong to the thrush family. Several species in the northern hemisphere are migratory, with a few breeding as far north as conditions permit and moving south for the winter. They include, among the larger thrushes, Redwings, Fieldfares, Dusky, Gray-cheeked, Siberian, and Varied thrushes, plus the smaller Northern Wheatear, Common Redstart, and Bluethroat.

Since the larger northern-breeding thrushes need cover for nesting and feeding, most do not occur in the true Arctic, where low-growing tundra plants predominate, but in the area to the south, a zone known in Eurasia as the "taiga" and in North America as "muskeg." Here scrub and scattered trees, often dotted with lakes, form a rich habitat where the insects and berries on which these birds feed abound.

Redwings breed throughout northern Europe and across northern Siberia almost to the Pacific coast. No population winters on the Pacific; all move south and west to reach a broad region running from the Atlantic, including western Europe and the Mediterranean, east to the Black and Caspian seas. The exception to this pattern of movement is the Icelandic population of Redwings. These birds, a different race, darker in color and with more heavily marked underparts, move south and east. Those breeding in eastern Iceland winter mainly in Britain and northern France; those from the western half of the island go farther south to southwestern France, Spain, and Portugal.

Both Redwings and Fieldfares are successful colonizers. Fieldfares have bred in small numbers in Iceland since the 1950s, and in Greenland, where their presence dates from a single influx in January 1937. A flock migrating south across the North Sea from Norway to Britain was blown westward. Individuals were seen in several parts of Greenland and those that found the milder areas in the island's southwest, where there are extensive stands of willow and birch scrub, survived the winter. Then, instead of migrating back to Europe, they stayed and bred, thereby founding a small colony of entirely sedentary birds, which is still there.

The most traveled North American thrush is the Gray-cheeked, which breeds in Arctic Canada, Alaska, and eastern Siberia north of the tree line, particularly along the banks of streams and lakes. Its migration takes it southeast through the eastern half of the U.S. and across the Caribbean to winter in northern South America.

BLUETHROATS (left) are beautifully marked small thrushes that breed widely in Europe and Asia, from the Arctic to the Mediterranean. Most do not, however, breed on the tundra since they need vegetation at least 3 feet (1 m) high for nesting and feeding. They also avoid areas of dense trees.

Most Bluethroats are migratory, many making long journeys, either to sub-Saharan Africa or across the Himalayas to India and Southeast Asia.

Fieldfare

THE SPEED OF FIELDFARE colonization can be rapid. These birds of Scandinavia and western Russia first nested as far south as Britain only in 1967; today there are perhaps a dozen pairs scattered in northern and eastern areas as far south as Kent. If this extensive range becomes established, a rapid expansion in numbers of breeding Fieldfares could take place.

The pattern of Fieldfare expansion has been interestingly different from that of the Redwing. A pair of Redwings first nested in Scotland in 1925; by 2000, there were well over 100 pairs, almost all still in Scotland.

REDWINGS EAT BERRIES and fruit, as well as invertebrates. Flocks of these large migrants, which need more food than smaller birds, can strip a hedge or tree of berries in a few days.

When the weather is severe in the wintering areas, the search for enough to eat can lead to conflict with large, aggressive residents, such as Mistle Thrushes. While stocks last, a single Mistle Thrush (or, often, a pair) sits on a well-berried tree and defends it against all comers.

At such times, windfall apples (above) and bird-feeder offerings can mean the difference between life and death for the migrants.

BLUETHROAT
Luscinia svecica
Wingspan
8–9 in / 20–23 cm
Weight
½–1 oz / 15–25 g
Journey length
1,600–3,750 miles /
2,500–6,000 km

REDWING
Turdus iliacus
Wingspan
13–14 in / 33–35 cm
Weight
1¼–2½ oz / 35–75 g
Journey length
600–4,000 miles /
1,000–6,500 km

FIELDFARE
Turdus pilaris
Wingspan
16–17 in / 39–42 cm
Weight
2½–4½ oz / 70–125 g
Journey length
600–3,000 miles /
1,000–5,000 km

S limit of
breeding range

Probably
resident in
Iceland and S tip
Greenland

N limit of
wintering range

20° N

0°

Buntings and finches

A quick change of feathers helps a speedy retreat from the Arctic winter's onset.

Few small birds breed in the Arctic. Because most small birds lose more heat than large ones, due to their proportionately greater body surface for their size, cold climates pose greater survival problems for them. In addition, most birds that breed in the brief Arctic summer must migrate south for the winter, often crossing large expanses of water on the way. Larger, longer-winged birds are better able to do this than smaller ones.

The most successful small birds breeding in the Arctic are three buntings—the Snow Bunting and Lapland and Smith's longspurs—and a number of finches, including the Redpoll and the Chaffinch. All but the Chaffinch, which is restricted to Europe and western Asia, breed right around the North Pole in the Arctic regions of North America, Europe, and Asia.

Northern Chaffinch populations breed beyond the Arctic Circle in Scandinavia and Russia. Their range in northern Russia has expanded enormously during this century as they have steadily colonized regions farther and farther north through the vast areas of spruce forest. They do not normally breed north of the tree line.

The Snow Bunting's breeding range extends northward to all the land and islands closest to the North Pole. Although migratory, Snow Buntings do not move to "traditional" winter homes and often stay as far north as winter conditions allow. When they do move farther south, to the eastern U.S. or Britain, for example, they haunt habitats that resemble the Arctic breeding grounds, such as bleak shores and open fields.

Lapland Longspur

Snow Bunting

THE SNOW BUNTING'S white or pale plumage helps it to survive cold conditions. Less heat is lost from a white surface than from a dark one, but, more importantly, white feathers are hollow and filled with air, which is an excellent insulator. In colored feathers, the internal spaces are filled with pigment cells: the more pigment, the darker the feather. Pigment cells conduct heat. Snow Buntings living in eastern Siberia, the coldest part of the Arctic, are whiter than those in relatively warmer areas, such as Iceland.

Summer

Winter

BIRDS ON MIGRATION are vulnerable to sudden changes in the weather. In March 1906, an estimated 1.5 million Lapland Longspurs, migrating north over Minnesota, were caught in a sudden heavy snowfall. In what became known as the "great bird shower," birds fell out of the sky and became hopelessly disorientated by streetlights. Some crashed into buildings and died; others froze to death.

Lapland Longspurs, like Redpolls and Snow Buntings, burrow into snowdrifts in spells of bad weather. Snow holes offer birds excellent protection from the wind that, by reducing the air temperature and disrupting the insulation provided by the feathers, can kill.

CHAFFINCHES (left) breeding in temperate latitudes are sedentary. In fact, there are records of birds that have moved only a few miles in a 10-year lifespan. Those breeding in northern latitudes are, of necessity, migratory, reaching Britain and Ireland in winter and frequently moving on south as far as France and Iberia.

Males and females migrate separately, so wintering flocks comprising birds of one gender are common. Females tend to move farther than males, with the result that males are more numerous in Sweden, the factor that prompted biologist Carl Linnaeus to give them the Latin name *coelebs*, meaning "bachelor."

CHAFFINCH
Fringilla coelebs
Wingspan
10–11½ in / 25–29 cm
Weight
²/₃–1 oz / 22–30 g
Journey length
0–3,200 miles /
0–5,000 km

SNOW BUNTING
Plectrophenax nivalis
Wingspan
11½–13 in / 29–33 cm
Weight
1–2 oz / 30–65 g
Journey length
0–3,800 miles /
0–6,000 km

LAPLAND LONGSPUR
Calcarius lapponicus
Wingspan
9½–11½ in / 24–29 cm
Weight
1–1½ oz / 30–40 g
Journey length
650–4,600 miles /
1,000–6,500 km

A unique adaptation of Snow Buntings to the rigors of life so far north is their rapid molt after breeding (see p. 24). They must complete their annual change of feathers quickly so that they have a new set, particularly of wing feathers, before the onset of winter and the need to migrate. Most small birds drop one or two wing feathers at a time and wait until the new ones are nearly grown before shedding the next. In this way, although flight is affected, it remains possible. Snow Buntings are in such a hurry that they lose as many as four or five (about half) of their main wing feathers at once. This means that they cannot fly for a week or two while the

new ones grow. For a ground-feeding bird like the Snow Bunting that lives mainly on seeds in late summer, in an area where predators are few and far between, this loss is a small price to pay.

All these small Arctic birds lay an average of six eggs, one or two more than similar species farther south. The Arctic birds are able to rear more young because the continuous daylight and abundance of insects at high latitudes allow them to feed their chicks almost around the clock. In fact, they do take a rest period of two or three hours in every 24 hours, and this is always in the hours around midnight, even though the sun may be shining.

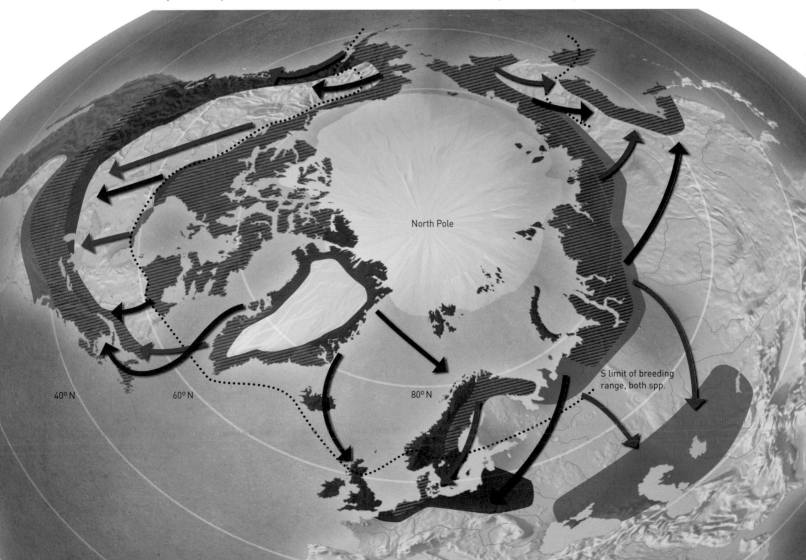

Southern hemisphere migrants

Patterns of bird movement in South America, Africa, and Australasia differ from those in the northern hemisphere. Because there is proportionately far less land in the southern hemisphere, there are also fewer temperate zones, the very areas that in the north are the breeding grounds for most migrants. Also, fewer birds have been banded in the south than in the north, so their movements are less well documented.

  With the exception of sea birds, which abound in the southern oceans, the birds that breed in the southern hemisphere fall into three main categories. Tropical and subtropical breeders usually have no need to move or migrate to take advantage of the burgeoning plants and insects that follow the rainy season. Many birds that breed in temperate latitudes migrate to spend the southern winter in warmer climates. Finally, altitudinal migrants spend most of their lives in the mountains, but move from higher elevations to lower ones after breeding.

Carmine Bee-eaters breed in colonies of up 10,000 pairs in the northern and southern tropics of Africa and migrate toward the equator in winter.

South American migrants ①

The Andes Mountains and Amazon River basin dominate migratory patterns in South America.

The composition of the bird life of South America is dominated by the abundance of species occupying the vast expanses of rainforest that cover almost all of the northern, broadest part of the continent. The equator crosses the continent at the Amazon delta. A wide variety of tropical and subtropical habitats extends across the huge Amazon basin and the extensive equatorial mountains of the northern Andes. An enormous diversity of bird species inhabits these high-rainfall regions. Many of the millions of migrants from North America that winter on the continent do so in this northern area.

The great length of the Andes, which stretch through 80° of latitude, means that there is a wide variety of plant and animal life associated with the mountains. A small proportion of the continent is arid, with true desert limited to the Pacific coasts of Chile, Peru, and southern Ecuador, and across the Andes into Argentina. These deserts include some of the driest places on earth—rainfall has never been recorded at Calama, Chile, in the Atacama desert. The southern tip of the continent is narrow, with the result that an extensive temperate fauna has not developed.

Migration is not a prominent feature of South America's land birds. The Amazon rainforest and northern tropics provide essentially stable conditions in which most birds are resident. The narrowness of the southern portion of the continent greatly limits the high-latitude land that, in the northern hemisphere, is extensive and provides the breeding habitat for most of the continents' migrants.

South American birds of temperate and high-altitude environments occur mainly along the Andes. For many of these species, some altitudinal migration is strongly suspected but not well documented. It is thought that birds nesting in the high mountains move to lowland regions after breeding.

Some South American land birds leave the continent in winter. Others breed in the south and then move north to the extensive equatorial regions. A few species are restricted in distribution to the temperate southern latitudes, with some migrating north in winter. There are no migrants among the breeding land birds of the Falkland and Galápagos islands.

✪ HOT SPOTS

This list includes places where wintering birds from North America, in addition to native Central and South American birds, both migratory and nonmigratory, may be seen.

① Mazatlán/San Blas, Mexico
② Rio Lagartos, Mexico
③ Belize City
④ Panama
⑤ Los Olivitos, Venezuela
⑥ Surinam coast
⑦ Paracas, Peru
⑧ Lagoa do Peixe, Brazil
⑨ La Serena, Chile
⑩ San Clemente del Tuyú/Cabo San Antonio, Argentina
⑪ Punta Rasa, Argentina
⑫ Tierra del Fuego

SOUTH AMERICA has two outstanding natural features that dominate its character. The first is the Amazon, the world's greatest river, which extends 4,000 miles (6,500 km) and has a volume of flow exceeding any other. The Amazon basin contains the largest single area of rainforest on earth.

A second superlative is provided by the Andes. These mountains hug the full length of the west coast and constitute the longest mountain chain in the world, with some of the highest peaks and many active volcanoes. The mountains form a barrier that divides the continent's drainage between a generally narrow region to the west and a much broader area to the east. Several major rivers, including the Orinoco and Paraguay–Paraná, drain this eastern region.

Many birds breed and winter in the savanna of the upper Orinoco basin from southwestern Venezuela to Colombia, and in the more extensive savanna of the central south from Brazil to Patagonia.

Major route

BLUE-AND-WHITE SWALLOW
Notiochelidon cyanoleuca
Wingspan
8½–10 in / 22–25 cm
Weight
⅓–½ oz / 10–15 g
Journey length
0–5,000 miles /
0–8,000 km

DARK-FACED GROUND TYRANT
Muscisaxicola macloviana
Wingspan
9–10½ in / 23–27 cm
Weight
⅗–⅘ oz / 18–25 g
Journey length
0–3,000 miles /
0–5,000 km

GIANT HUMMINGBIRD
Patagona gigas
Wingspan
7½–8½ in / 19–21 cm
Weight
⅗–⅔ oz / 17–20 g
Journey length
0–500 miles /
0–800 km

South to north migrations

Most of the well-understood migrations of land birds in South America fall into this category, with birds breeding in the temperate areas of the south and moving north for the southern winter.

Some of these birds undertake spectacular migratory journeys from the extreme south as far as, and sometimes even across, the equator. They include the Blue-and-white Swallow and the Brown-chested Martin, which breeds as far south as southern Chile and Argentina and occasionally reaches as far north as Mexico on migration.

Others, however, travel less far, breeding in the temperate regions of central Argentina and Chile and migrating no farther than northern South America. Such birds include a number of tyrant flycatchers, members of one of the world's largest bird families, found only in the New World. It is South America's dominant family of land birds in terms of number of species, constituting about 23 percent of all perching birds in some regions. More than 70 species have been counted in some areas of western Amazonia alone. Most of the species breeding in temperate areas are migratory, including the Dark-faced Ground Tyrant and the Fork-tailed Flycatcher.

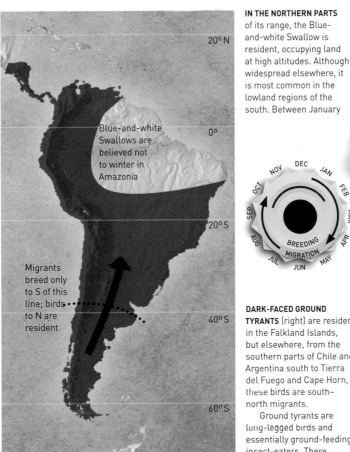

IN THE NORTHERN PARTS of its range, the Blue-and-white Swallow is resident, occupying land at high altitudes. Although widespread elsewhere, it is most common in the lowland regions of the south. Between January and March, when the breeding season is over, these birds fly north, many following the line of the Andes. The return flight in spring takes place in August.

Blue-and-white Swallow

Giant Hummingbird

THE LARGEST OF ALL THE HUMMINGBIRDS, the Giant Hummingbird (left) is 10 times larger than the smallest member of this family of specialized nectar-sipping birds. Giant Hummingbirds are mountain birds, occurring in the Andes from Ecuador in the north to Chile and Argentina in the south. Populations from the south of the range are migratory.

DARK-FACED GROUND TYRANTS (right) are resident in the Falkland Islands, but elsewhere, from the southern parts of Chile and Argentina south to Tierra del Fuego and Cape Horn, these birds are south–north migrants.

Ground tyrants are long-legged birds and essentially ground-feeding insect-eaters. There are nine other species of this group in South America, all of which occur along the Andes chain and all of which are highly migratory.

The "tyrant" part of the family name derives from the birds' aggressive behavior toward other species, including birds of prey, especially during the breeding season.

South American migrants ②

KELP GOOSE
Chloephaga hybrida
Wingspan
3 ft 11 in–4 ft 7 in / 1.2–1.4 m
Weight
4 lb 7 oz–5 lb 12 oz / 2–2.6 kg
Journey length
0–500 miles /
0–800 km

Migrations within temperate latitudes

Few birds of the temperate southern forests and coastal lowlands of South America have clearly defined northward migrations in winter. Of those that do, the Kelp Goose and Two-banded Plover are typical, migrating short distances to join resident populations farther north.

The Black-faced Ibis also falls into this group, breeding in southern Chile and Argentina and wintering on the pampas of northern coastal Argentina. There are also sedentary populations of this species in the highlands of Ecuador, Peru, Bolivia, and northern areas of Chile.

THERE ARE FIVE SPECIES of gooselike birds in South America, all of which are grazers.

The Kelp Goose (right) is unusual among them in that it is confined to maritime areas, haunting rocky coasts and feeding on seaweed. Birds that breed in the south of the range, in Tierra del Fuego, fly north in winter but stay along the shorelines. Kelp Geese in the Falkland Islands are resident.

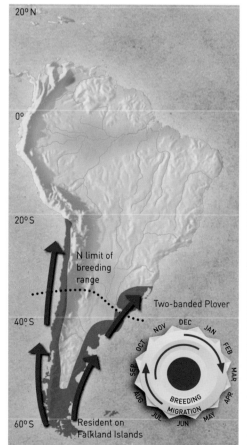

Map labels: 20° N — 0° — 20° S — 40° S — 60° S — N limit of breeding range — Two-banded Plover — Resident on Falkland Islands — NOV DEC JAN FEB MAR APR MAY JUN JUL AUG SEP OCT — BREEDING — MIGRATION

THE TWO-BANDED PLOVER (above) occurs along most shores and coasts of the continent south of the Tropic of Capricorn. Its favorite haunts are beaches, estuaries, and river banks near the sea.

Falklands Islands birds are resident, but elsewhere populations are migratory, moving northward for the southern winter to reach southern Chile in the west and Uruguay in the east.

TWO-BANDED PLOVER
Charadrius falklandicus
Wingspan
18–22 in / 45–55 cm
Weight
1²/₅–2 oz / 40–60 g
Journey length
0–2,250 miles /
0–3,600 km

ANDEAN GOOSE
Chloephaga melanoptera
Wingspan
4 ft 7 in–5 ft 3 in / 1.4–1.6 m
Weight
5 lb 8 oz–7 lb 11 oz /
2.5–3.5 kg
Journey length
0–3 miles / 0–5 km

GRAY GULL
Larus modestus
Wingspan
3 ft 3 in–3 ft 7 in / 1–1.1 m
Weight
5 lb 8 oz–7 lb 11 oz /
2.5–3.5 kg
Journey length
20–1,600 miles / 30–2,500 km

Andean Geese breed at high altitudes and winter lower down

Andean Goose

Altitudinal migrants

The Andes provide a corridor of suitable habitat for temperate species colonizing the northern tropics. The mountains are also the route along which some long-distance migrants travel. And there are species that spend the summer at high altitudes in the Andes, then travel only short distances to reach the surrounding lowlands in winter.

There are few well-documented examples of altitudinal migration among South American birds, but the migration of the Andean Goose is believed to be typical of these altitudinal movements.

ANDEAN GEESE (left) breed high in the mountains, often along the shores of the Andean lakes, at altitudes from 10,000 feet (3,300 m) up to the snow line. Many remain at high altitudes year-round; others are migratory.

Although they breed in isolated pairs, flocks usually come together outside the breeding season, as the birds graze the marshy plains of the Andean foothills and valley bottoms.

While males and females are similar in appearance, males are substantially larger.

Desert breeders

Few birds breed in the harsh conditions of the South American deserts. The Gray Gull is an exception, not only migrating to coastal regions at the end of the breeding season but also, even while nesting, "commuting" daily from its high-altitude nest to the coastal lowlands to feed.

Non-breeding

Gray Gull

AT DUSK, Gray Gulls leave coastal areas to return to their inland nests. These birds breed in areas where there is no rainfall—colonies have been found in several locations in the Atacama Desert, where the Andes rise steeply from the sea.

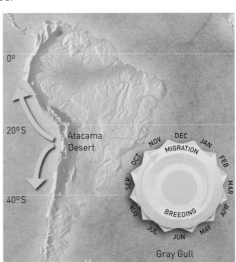

Gray Gull

African migrants ①

Migration patterns in the largest southern continent are governed by complex climatic variables.

When the glaciers that covered much of Europe began to retreat at the end of the last ice age, new tracts of land became available for habitation and colonization by plants and animals. Most of the species that spread into Europe were from Africa, the largest southern continent. The kinds of birds that inhabit Europe today are the result of the continent's proximity to the African landmass.

When the ice age ended 10,000 years ago, the Sahara was not desert but woodland. Today this broad, arid wasteland is treated as part of the Palearctic faunal zone, along with the temperate and Arctic areas of Eurasia. All of Africa south of the Sahara desert is called the Afrotropical zone.

Many bird families occur in both Europe and Africa, but those that contain but a single species in Europe—such as the cuckoos, kingfishers, nightjars, and bee-eaters—can boast 10 or more species in Africa. In addition to providing a wintering area for hundreds of millions of migrant birds from Eurasia, the Afrotropical zone supports a large body of intra-African migrants. Many of these have extensive breeding ranges throughout the region, but those that extend to the southern portion of the continent breed in the southern summer (October to March) and fly northward to winter at lower latitudes, some crossing the equator. Species breeding north of the equator breed in the northern summer, then also move to lower latitudes. They are fewer in number because the belt of savanna woodland is narrower in the north than in the south.

In the equatorial region between 10°N and 10°S, the terms "summer" and "winter" become meaningless—the day length of some 12 hours varies little between the summer and winter solstices and there are no prolonged periods of twilight at dawn and dusk. The seasons are instead determined by intervals of wet and dry weather. Close to the equator, there are two rainy periods, a short one and a long one, interspersed with dry periods when rainfall is greatly reduced or rare.

With increasing distance north and south of the equator, the year becomes progressively more evenly divided between a warm rainy period (during which most birds breed) and a cool dry period, when the most northerly and southerly breeders may migrate.

➡ **Major route**

✪ HOT SPOTS

Native African birds or Eurasian visitors to the continent, or both, may be seen at these sites.

① Gulf of Suez/Nile delta, Egypt
② Oued Massa, Morocco
③ Banc d'Arguin NP, Mauritania
④ Senegal and Gambia rivers, Senegal/Gambia
⑤ Niger River basin, Mali
⑥ Air and Ténéré NNP, Niger
⑦ Lake Chad, Chad/Niger
⑧ Mount Ngulia, Kenya
⑨ Luangwa Valley NP, Zambia
⑩ Victoria Falls NP, Zambia/Zimbabwe
⑪ Moremi WR, Botswana
⑫ Nylsvley NR, South Africa
⑬ Langebaan Lagoon, South Africa

THE LANDSCAPE OF AFRICA is dominated by rift valleys and high mountains. The continent's highest mountains—Kilimanjaro and Kenya—are, however, isolated peaks with the result that they are rarely barriers to migration. Indeed, some birds seem to use them as stepping stones on their travels. The Sahara desert, which covers an area of 3.3 million sq miles (8.6 million km²), is by contrast a formidable barrier for many migrants journeying to and from Africa

The continent's major rivers—the Nile, Zaïre, and Niger—are important pathways for migrants, since their flanking vegetation affords food and somewhere to roost, particularly in the dry savanna regions. The chains of lakes and rift valleys, dominated by the Great Rift Valley in the east, provide attractive routes and destinations for many shorebirds, while their large reedbeds offer areas in which swallows and martins find plentiful roosts.

AFRICAN PYGMY KINGFISHER
Ceyx picta
Wingspan
6¼ in / 16 cm
Weight
½ oz / 15 g
Journey length
600–1,250 miles /
1,000–2,000 km

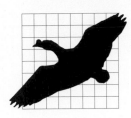

AFRICAN COMB DUCK
Sarkidiornis melanotos
Wingspan
3 ft 8 in–5 ft 2 in / 1.1–1.4 m
Weight
4 lb 6 oz / 2 kg
Journey length
2,200–2,400 miles /
3,500–3,900 km

Mid-continent migrations

The areas north and south of the equator may shelter many migrant species. It seems likely that the most northerly and southerly breeders in this area may migrate to lower latitudes. However, because many of these species appear to be "resident" breeders over very large areas of central Africa, it is seldom possible to delineate the breeding and nonbreeding ranges of the different populations. Outside the breeding season birds in any area may be residents, migrants, or a combination of the two. Pygmy Kingfishers are an exception to this generalization, since they are known migrants.

Migrating north of the equator

The migratory journeys of many African populations take them across the equator to the continent's northern belt of savanna woodland. These include representatives of many families, such as the Comb Duck, Carmine Bee-eater, Pennant-winged Nightjar, and Wahlberg's Eagle.

In March and April, around 100,000 Wahlberg's Eagles leave their breeding grounds in the southern woods and head north. Some stay south of the equator in southern Zaïre; others reach Ethiopia. The champion is a nestling banded in the Transvaal, South Africa, and recovered 2,600 miles (4,200 km) away in Sudan.

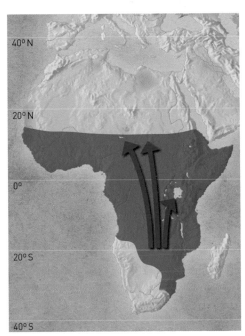

African Pygmy Kingfisher

PYGMY KINGFISHERS (top) breed in southern Africa and winter as far north as Zaïre and Mozambique. Similar movements appear to occur north of the equator. These birds are faithful to their breeding areas; one individual banded in the Mkume Reserve, Zululand, was recovered there five years later.

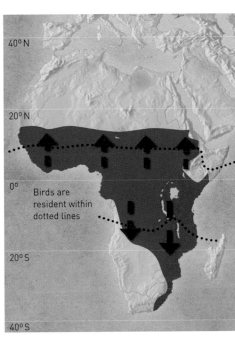

Birds are resident within dotted lines

Comb Ducks

COMB DUCKS occur in the tropics of South America and Asia, as well as in Africa where they range from The Gambia to Sudan and south to Natal, South Africa. These birds are partial migrants, so those present in an area may comprise residents, migrant visitors, or both.

Several Comb Ducks banded in Zimbabwe have been recovered in Chad and Sudan, up to 2,400 miles (3,900 km) away.

African migrants ②

THE CARMINE BEE-EATER (right) is the largest African bee-eater. These birds breed in colonies numbering several thousand individuals, crowding high river banks with their nest tunnel entrances.

There are two separate populations, one in the north, of blue-throated birds, and one in the south with a breeding nucleus in the Luangwa River valley in Zambia. These southern birds usually fly south at the end of the breeding season to southern Mozambique and the Transvaal. There they may be seen from December to March, before they move north again, some as far as the equator.

Carmine Bee-eaters are often nomadic between breeding seasons and are strongly attracted by grass fires (an ever-present feature of the African dry season). The birds wheel through the smoke to prey on the flying insects fleeing the flames.

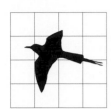

CARMINE BEE-EATER
Merops nubicus
Wingspan
18–20 in / 45–50 cm
Weight
2 oz / 60 g
Journey length
300–750 miles /
500–1,200 km

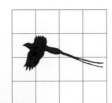

**AFRICAN PARADISE
FLYCATCHER**
Terpsiphone viridis
Wingspan
9½–11 in / 24–28 cm
Weight ½ oz / 15 g
Journey length
300–1,100 miles /
500–1,800 km

**SOUTH AFRICAN
CLIFF SWALLOW**
Hirundo spilodera
Wingspan
10¼–11½ in / 26–29 cm
Weight ¾ oz / 21 g
Journey length
1,250–1,550 miles /
2,000–2,500 km

Migrations to the tropics

Few species in Africa have discrete southern breeding and equatorial wintering ranges. Among those that do are the African Paradise Flycatcher and Amethyst Starling. These birds—which winter in Zambia and Angola and on occasion cross the equator into Sudan—return year after year to the same breeding sites. Recoveries of Cliff Swallows, which breed in South Africa, have demonstrated that they winter in western Zaïre, a journey of some 1,500 miles (2,400 km) which they accomplish in around two weeks.

Other African birds whose migrations fall into this category include the Red-breasted Swallow and the African Reed Warbler.

BANDING RECOVERY RATES of small birds are low, rarely exceeding 1 in 300 (see pp. 42–43). This makes more remarkable the recovery of two Paradise Flycatchers banded in South Africa. One, banded near Pietermaritzburg in Natal province, was next noted in the Nambuli Mountains of Mozambique. The other, banded near Rustenburg in the Transvaal, was recovered in Zimbabwe during October, evidently on its way back to the breeding grounds and still some 350 miles (550 km) north of its target.

African Paradise Flycatcher

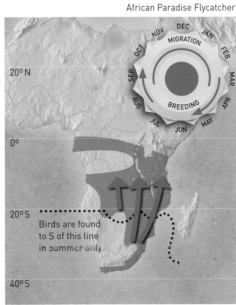

Birds are found to S of this line in summer only

BETWEEN SEPTEMBER AND MARCH, Cliff Swallows (above) breed in large colonies under bridges and in culverts in the South African highveld. Nets dropped over both ends of a culvert make it easy to capture and band an entire colony.

Banding has shown that some of these birds are remarkable survivors, living to be eight or nine years old.

South African Cliff Swallow

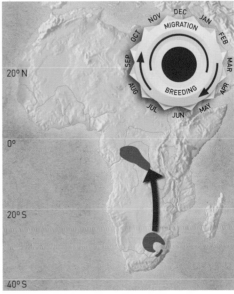

African migrants ③

Altitudinal migrants

Mountains and escarpments characterize the eastern and southern parts of Africa and provide a unique and important habitat for many species of flora and fauna. Among the birds that inhabit these mountains are many that leave the cloudy forests and rocky heights at the end of the rainy season to move to lower elevations. Altitudinal migration frequently occurs over short distances, with many species reaching only the surrounding foothills.

0°

Birds breed and winter at different elevations

20° S

Starred Robin

NOV DEC JAN
OCT   FEB
MIGRATION
SEP   MAR
AUG   APR
BREEDING
JUL JUN MAY

STARRED ROBIN
Pogonocichla stellata
Wingspan
9–11 in / 23–27 cm
Weight
¾ oz / 21 g
Journey length
6–120 miles /
10–200 km

GREATER CRESTED TERN
Thalasseus bergii
Wingspan
3 ft 10 in–4 ft 1 in /
1.18–1.25 cm
Weight 11 oz / 300 g
Journey length
600–1,000 miles /
1,000–1,600 km

STARRED ROBINS (right) are altitudinal migrants, but unusual in that they also make migrations of 60 miles (100 km) or more in the dry season into hot river valleys where they move through the narrow strips of evergreen growth along the dry riverbeds.

Littoral migrants

The coastlines and inshore waters of Africa shelter many native marine birds. Some, such as the Greater Crested Tern, are littoral migrants, breeding and wintering on discrete stretches of shoreline. The Damara Tern is alone in breeding on the inhospitable Skeleton Coast of Namibia. At the end of the breeding season, these birds fly north to "winter" in the Gulf of Guinea. There, they avoid competition for resources with migrant terns from Eurasia by feeding well out to sea.

Greater Crested Tern

NOV DEC JAN
OCT   FEB
SEP   MAR
BREEDING
AUG MIGRATION APR
JUL JUN MAY

○ Breeding areas

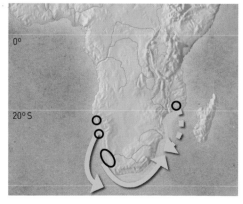

0°

20° S

IN THE SOUTH, Crested Terns (right) nest on small rocky islands from Namibia down to Dyer Island, northwest of Cape Agulhas, the most southerly point of Africa. Once the young birds can fly, they move southward and eastward along the coast as far as southern Mozambique.

Australasian migrants ①

The smallest and most remote continent has a wealth of bird species but few migrants.

Until the Torres Strait was flooded some 10,000 years ago, Australia had a broad land connection north to New Guinea. As a consequence, although many Australian plant and animal species share close affinities with species found in New Guinea and the islands of the Indonesian archipelago, the distinctive flora and fauna of Australia evolved in isolation from all other parts of the world.

The New Zealand island group has long been remote from other centers of evolution. Its flora and fauna reflect this fact in the relatively small number of different life forms and the presence of some unique species, including the existence there of forms that have long since died out elsewhere. The isolation of these islands meant that many species evolved in a forested environment free from predatory mammals. Several of the land birds became highly specialized and many lost their powers of flight.

Fewer than 15 percent of the indigenous land birds of Australia and New Zealand are judged to be truly migratory. Over much of the Australian mainland the survival of birds depends on their ability to cope with arid environments. In these areas conditions are rarely consistent from season to season, and a flexibility in movement patterns of bird populations is often the preferred strategy for survival.

Such movements cannot be classed as migratory because they are neither regular in annual occurrence nor predictable in direction of dispersal. Birds simply assemble and breed in a favorable region and depart when conditions deteriorate. There are many essentially sedentary species in both the temperate and tropical forests of Australia and even a few species that remain within some of the arid and semi-arid regions year-round.

Nor is migration common among New Zealand's land birds, although there is some internal migration, with species breeding in South Island and wintering in North Island. Altitudinal movements occur irregularly, in response to severe weather in the mountains. The country's sea and coastal birds, some 40 percent of the total avifauna, may by contrast wander considerable distances from their breeding areas outside the breeding season.

⊗ HOT SPOTS

1. Gulf of Carpentaria, Northern Territory
2. Arnhem Land, N Territory
3. Eighty Mile Beach/Roebuck Bay, W Australia
4. Port Hedland, W Australia
5. Shark Bay/Lake Macleod, W Australia
6. Lake Eyre, S Australia
7. St. Vincent Gulf/The Coorong, S Australia
8. Port Philip Bay/Corner Inlet, Victoria
9. Miranda, North Island, New Zealand

THERE ARE THREE broad climatic zones in Australia. Parts of Western Australia, Northern Territory, and Queensland are tropical. Tasmania and the higher areas of the continent's southeast are temperate. Most of the remainder of Australia is arid and semi-arid. In addition, the coastal southeast and southwest are "Mediterranean," with cool wet winters and hot dry summers. The mountains of the Great Divide separate the narrow coastal plains to the east from the dry interior. New Zealand is mountainous, with high rainfall.

These two major landmasses are surrounded by many offshore groups of smaller islands, ranging northward to the subtropics and southward to the subantarctic.

Major route ➡

Australasian migrants ②

Migrants from Tasmania to Australia

Several birds breed in Tasmania and winter in Australia. They include the Orange-bellied Parrot—which travels from the island's southwestern coast to the dunes and foreshores between The Coorong and Corner Inlet—and the Swift Parrot.

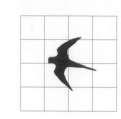

SWIFT PARROT
Lathamus discolor
Wingspan
13–14½ in/33–36 cm
Weight
1¾–2⅔ oz/50–74 g
Journey length
200–1,550 miles/
350–2,500 km

Migrations north from New Zealand

The Double-banded Plover is one of the few land birds migrating between New Zealand and Australia, with some populations regularly crossing the Tasman Sea. This group also includes the Long-tailed Cuckoo, which breeds in New Zealand and winters on the tropical islands of the Pacific, and the Shining-bronze Cuckoo.

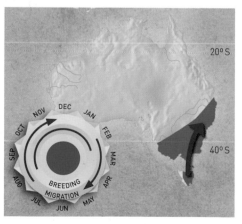

KNOWN TO BREED only on the island of Tasmania, the Swift Parrot (above) regularly migrates to the eastern Australian mainland in winter.

The birds occur widely, but in scattered locations, in the forested areas of the east coast as far north as southern Queensland. But their chattering, flutelike calls are most commonly heard in the wooded regions of southern Victoria as flocks forage on the blossom-laden eucalyptus trees.

GOLDEN-BRONZE CUCKOO
Chalcites lucidus
Wingspan
12–14 in / 30–34 cm
Weight
1–1½ oz / 28–45 g
Journey length
200–3,400 miles /
350–5,500 km

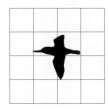

SACRED KINGFISHER
Halcyon sancta
Wingspan
11½–13½ in / 29–33 cm
Weight
1–2 oz / 28–58 g
Journey length
0–2,400 miles /
0–3,900 km

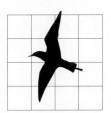

AUSTRALIAN BEE-EATER
Merops ornatus
Wingspan
16–18 in / 40–45 cm
Weight
⅔–1⅛ oz / 20–33 g
Journey length
0–3,000 miles /
0–4,800 km

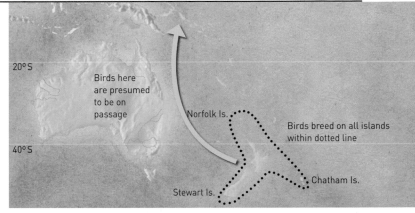

AUSTRALIAN GOLDEN-BRONZE CUCKOOS (left) migrate north as far as New Guinea, the Lesser Sunda Islands, and the Bismarck Archipelago.

The New Zealand Shining-bronze Cuckoo, once regarded as a separate species from these Australian birds, is now classed as the same species. New Zealand birds breed throughout the island group, south to Stewart Island, east to the Chatham Islands, and as far north as Norfolk Island.

Their migratory route in the fall is not well understood and may involve a direct flight north, at least for the adults. Or, as is believed to be the case in spring, birds may hug—or even stop over for some time on—the east coast of Australia.

Australian Golden-bronze Cuckoo

Birds here are presumed to be on passage

Norfolk Is.

Birds breed on all islands within dotted line

Stewart Is.

Chatham Is.

Migrants from Australia to the tropics

Several land birds migrate regularly between the continent of Australia and the tropical islands to the north, in particular to New Guinea and the surrounding area. Among the better known of these migrants are the Channel-billed Cuckoo, Sacred Kingfisher, and Australian Bee-eater.

Populations that breed in the north of the continent are often resident or only locally dispersive. Those that breed in the southern areas, however, must move north to avoid the less favorable conditions of the southern winter.

SACRED KINGFISHERS (right) are widely distributed throughout Australia. Northern populations are resident, but southern birds, which breed in all but the driest areas, join the residents or fly on to Sumatra, Borneo, Sulawesi, New Guinea, and the surrounding islands in winter.

New Zealand populations are also resident, but some are altitudinal migrants, moving in winter from the mountains to nearby lowlands.

Australian Bee-eater

AUSTRALIAN BEE-EATERS breed in all parts of Australia except Tasmania, with southern populations migrating northward to winter in northern Australia, eastern Indonesia, and New Guinea.

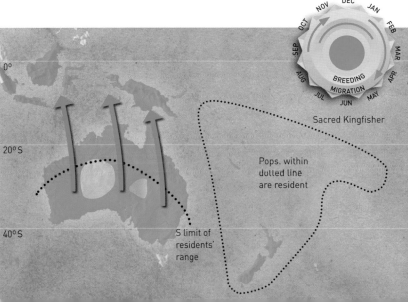

Sacred Kingfisher

Pops. within dotted line are resident

S limit of residents' range

Australasian migrants ③

Migrations within Australasia

Species that migrate altitudinally or from southern regions northward each fall and return in spring include the majority of the true migrants in Australasia. Among this group are several of the honeyeaters, such as the Little and Noisy friarbirds and the Scarlet and Yellow-faced honeyeaters. The fall migration of the Yellow-faced Honeyeater is most apparent in the Canberra area as mixed flocks begin to leave the high-altitude forests. Despite this conspicuous start, details of the whole migration are unclear and at least some birds stay in the south for the winter.

Migrants within New Zealand include the South Island Pied Oystercatcher, which breeds in the river valleys of the south and winters on the estuaries of North Island.

SCARLET HONEYEATER
Myzomela sanguinolenta
Wingspan
9–10 in / 22–25 cm
Weight
⅓–½ oz / 12–15 g
Journey length
30–1,550 miles /
50–2,500 km

SPECTACULAR MALE SCARLET HONEYEATERS
(left) can often be detected in spring by their tinkling song as they forage on eucalyptus and melaleuca blossom. Although they occur along Australia's eastern seaboard, these birds are also known on islands to the north from Sulawesi to New Caledonia. On the islands, however, they are sedentary, and the existence of separate races here suggests that there is little dispersal, and indeed considerable isolation, of the different populations.

In Australia, Scarlet Honeyeaters breed in the southern forests and generally move north to the coastal forests in winter.

0°

NOV DEC JAN
OCT MIGRATION FEB
SEP MAR
AUG APR
JUL BREEDING MAY
JUN

Scarlet Honeyeater

Pops. within dotted line are resident

20° S

Resident here

Birds to N of dotted line are resident; birds breeding to S migrate N in winter

40° S

LITTLE FRIARBIRD
Philemon citreogularis
Wingspan
15–19 in / 38–48 cm
Weight
2²⁄₃–8 oz / 75–220 g
Journey length
30–1,500 miles /
50–2,400 km

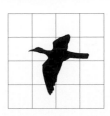
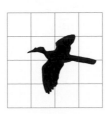

NOISY FRIARBIRD
Philemon corniculatus
Wingspan
19–22 in / 48–55 cm
Weight
5¼–8 oz / 150–220 g
Journey length
30–1,000 miles /
50–1,600 km

GRAY-BACKED WHITE-EYE
Zosterops lateralis
Wingspan
13–14½ in / 33–36 cm
Weight
½–²⁄₃ oz / 15–20 g
Journey length
30–1,250 miles /
50–2,000 km

SOUTHERN BREEDING LITTLE FRIARBIRDS (left) are migratory; farther north birds are more nomadic, often staying in an area while the blossoms last, then moving on.

This species also occurs in New Guinea, although there is no evidence of movement between populations.

CLASSED AS TRUE MIGRANTS in the south of their eastern Australian range, Noisy Friarbirds are more nomadic in the north. They also occur in New Guinea.

These honeyeaters are large, conspicuous, and somewhat ugly-looking birds, best known for their distinctive, raucous calls as they forage in the upper canopy of tall eucalyptus groves.

Noisy Friarbird

THE GRAY-BACKED WHITE-EYE (left) is widely distributed in Australia and is a recent colonist in New Zealand. It has also spread to many islands in the Pacific as far north as Fiji. In Australia it occurs around the coasts from the Tropic of Capricorn to Tasmania and north to Cape York.

The populations of southeastern Australia are spectacular migrants, moving north in winter, either toward South Australia or up the east coast as far as Queensland. Banded Tasmanian breeders have been recovered in northern Queensland.

Migratory sea birds

Nearly three-quarters of the Earth's surface is ocean, the home of sea birds. Although they must breed on land, sea birds spend most of their lives far out to sea, often moving long distances between seasons, not just over one ocean, but sometimes flying between them. Some sea birds breed and winter within comparatively small areas, but many are long-distance migrants, traveling thousands of miles on journeys that can take them from the far north of the northern hemisphere to the limits of the Antarctic pack ice in the southern hemisphere.

Little more than 300 of the 9,000 or so known bird species are sea birds. Yet they occur throughout the world and are the only group of birds to have successfully colonized Antarctica, the most inhospitable continent. And, although the number of species is small, many of their populations number in the millions. There are perhaps five million pairs of Chinstrap Penguins nesting on the South Sandwich Islands alone, and several colonies of Dovekies in the northern hemisphere exceed one million pairs. The presence of such huge numbers depends on the abundant food in the surrounding seas.

The Black-browed Albatross and other birds of the southern oceans may circumnavigate the globe between breeding seasons.

Patterns of migration

Food for sea birds accumulates where warm and cool ocean currents meet.

The sea is not a uniform habitat. There are immense variations between different parts, particularly in the amount of plankton and numbers of fish that live there. Sea birds feed either from the surface of the sea or in the top 100–170 feet (30–50 m) and must find areas where this thin layer contains plenty of food. This they do either where the water is comparatively shallow, such as close to coasts, or where the meeting of cold and warm currents produces turbulence in the form of upwellings of cold water. The highest density of plankton—the smallest, most primitive life form in the oceans—occurs in these regions, attracting squid and fish and, in turn, seals, whales, and birds.

By far the most important area of upwellings for all sea birds, and one that explains the enormous numbers of birds that breed in Antarctica, lies close to the Antarctic pack ice in an area known as the Antarctic Divergence. Here, a warm current that has flowed south from the Atlantic, Pacific, and Indian oceans rises to the surface, bringing with it abundant nutrients. These stimulate huge "blooms," or growths, of phytoplankton, minute plants that flourish in the continuous daylight of the Antarctic summer. They are fed on by animal plankton that in turn feed the krill, swarms of small shrimplike organisms whose numbers can exceed 30,000 in a cubic yard (39,000 in a cubic meter) of water. Krill can sustain all the larger marine life of the region, from fish to whales, and every sort of sea bird, whether they feed directly on the krill or on the abundant fish species higher in the food chain.

Plankton flourishes in cold water. The warm waters of the current are cooled by contact with the pack ice and flow north again, carrying the plankton in their stream. The cold water sits on top of the warm because it is mixed with substantial quantities of fresh water (from the summer meltwaters of the ice cap and from glaciers), which is lighter than salt water. It is not until this northward flowing cool water has traveled several hundred miles that it sinks below much warmer surface water in a region known as the Antarctic Convergence. There is thus a broad band of cold water around Antarctica that is rich in food and home not only to the millions of sea birds breeding in the region, but also to many

sea birds that breed in the northern hemisphere and migrate here for the northern winter.

The same conditions of nutrient-rich water and abundant plankton and fish also occur elsewhere. Major currents flow

THE CENTRAL REGIONS of the major oceans—the Pacific, Atlantic, and Indian—all lie in tropical or subtropical latitudes where temperatures are more or less constantly warm. Here there is little plankton growth and comparatively low densities of fish, with the result that these areas tend to be less attractive to sea birds than areas of turbulence. This, combined with the presence of the rich coastal currents, channels much sea bird migration (which is often considered—wrongly—to be on a broad front, across the widest stretches of sea) into comparatively narrow bands, which follow quite closely the coasts of the continents. Sea birds moving south from the North Atlantic, for example, tend to stay close to the western coasts of Europe and Africa, where the Benguela Current offers ample feeding opportunities. Some, however, do make the relatively short ocean crossing from western Africa to northeastern South America before continuing down the coast to Cape Horn and beyond.

MANY SEA BIRD species roam vast distances from their breeding colonies, traveling the oceans in search of nutrient-rich waters where food is plentiful. Several species cross the equator on migration.

WITH A BREEDING POPULATION estimated at hundreds of millions, Wilson's Storm-Petrel is probably the world's most numerous sea bird. Small in size and weighing only 1⅛ ounce (32 g), it travels from Antarctica as far north as Labrador.

SABINE'S GULLS are long-distance migrants. Those breeding in the marshy tundra and islets of Arctic Alaska and Siberia winter off the coasts of Peru and Colombia; Canadian and Greenland breeders migrate to southern Africa.

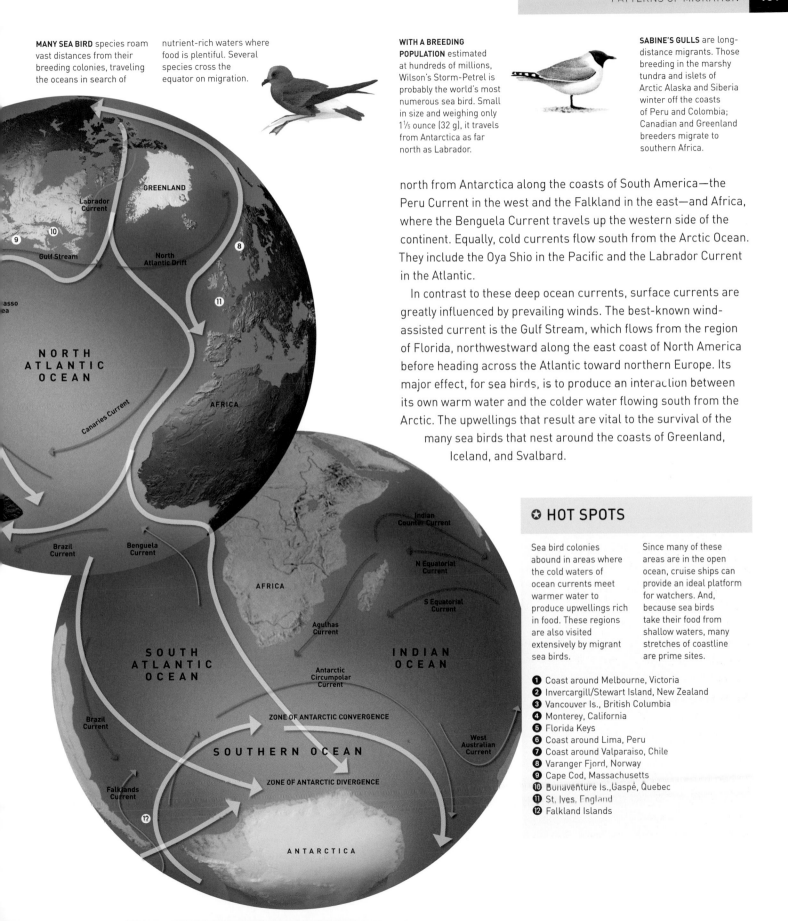

north from Antarctica along the coasts of South America—the Peru Current in the west and the Falkland in the east—and Africa, where the Benguela Current travels up the western side of the continent. Equally, cold currents flow south from the Arctic Ocean. They include the Oya Shio in the Pacific and the Labrador Current in the Atlantic.

In contrast to these deep ocean currents, surface currents are greatly influenced by prevailing winds. The best-known wind-assisted current is the Gulf Stream, which flows from the region of Florida, northwestward along the east coast of North America before heading across the Atlantic toward northern Europe. Its major effect, for sea birds, is to produce an interaction between its own warm water and the colder water flowing south from the Arctic. The upwellings that result are vital to the survival of the many sea birds that nest around the coasts of Greenland, Iceland, and Svalbard.

✪ HOT SPOTS

Sea bird colonies abound in areas where the cold waters of ocean currents meet warmer water to produce upwellings rich in food. These regions are also visited extensively by migrant sea birds.

Since many of these areas are in the open ocean, cruise ships can provide an ideal platform for watchers. And, because sea birds take their food from shallow waters, many stretches of coastline are prime sites.

1. Coast around Melbourne, Victoria
2. Invercargill/Stewart Island, New Zealand
3. Vancouver Is., British Columbia
4. Monterey, California
5. Florida Keys
6. Coast around Lima, Peru
7. Coast around Valparaiso, Chile
8. Varanger Fjord, Norway
9. Cape Cod, Massachusetts
10. Bonaventure Is., Gaspé, Quebec
11. St. Ives, England
12. Falkland Islands

Penguins

Fasting males incubate their eggs in freezing blizzards as females feed at sea.

All species of penguins are confined to the southern hemisphere, although the Galápagos Penguin nests close to the equator and has wandered across it as far north as Panama. The majority of species, however, occur well to the south, with several nesting in Antarctica. Here, they are extremely numerous—colonies of hundreds of thousands are not uncommon.

The species that nest in the warmer waters around South America and southern Africa are largely sedentary, wandering only short distances out to sea when they are not ashore breeding. Those nesting on oceanic islands farther south tend to move north during the winter, following the fish that make up their diet. The Antarctic species, which need open water in which to feed, must stay beyond the northern limits of the pack ice.

Little is known about penguin movements, other than in the breeding season, because the birds are difficult to follow while at sea. The appearance of juveniles long distances from the nearest colony suggests that these birds wander widely, in particular immediately after they leave the colony. Young King and Little penguins can move more than 600 miles (1,000 km), while juvenile Jackass Penguins, which breed on the coasts of South Africa, may reach the equator on the continent's Atlantic seaboard.

In the breeding season, many Antarctic species make long, regular trips between nests and feeding grounds. King Penguins are the champions. Small young, which require frequent feeding, are guarded by one parent while the other may travel up to 250 miles (400 km) for food. Once the young are old enough to be left by both parents in a nursery, the feeding trips may double in length.

The Emperor Penguin is unique among Antarctic birds in that it breeds in winter. This is probably so that when the chick is ready to leave the colony, at about five months old, food is plentiful and conditions are favorable for its survival. As soon as she has laid the single egg, the female passes it to the male, who places it on top of his feet and lowers a pouch of skin over the top. The female then leaves the colony to feed, returning when the egg is ready to hatch.

The male incubates the egg for up to 65 days, keeping it at a temperature of 88°F (31°C); around him temperatures may fall to –76°F (–60°C) and winds blow at 125 mph (200 km/h). To conserve heat, the males huddle in dense packs, taking turns to shuffle into and out of the center of the group, where the temperature is several degrees higher than on the fringes.

Adélie Penguin

ALTHOUGH ADÉLIE PENGUINS rarely travel more than 30 miles (50 km) from their colonies to find food, they are capable of navigating over longer distances if necessary. Birds taken from the Antarctic coast and released inland on featureless snow fields headed north, the direction that would inevitably bring them to the sea and thus to food and safety. That these movements were confined to periods when the birds could see the sun proved that they were navigating with its aid.

MAGELLANIC PENGUINS breed on the Falkland Islands and along the coasts of Argentina and Chile. They nest in colonies of up to 200,000 pairs, laying their eggs in burrows or shaded depressions that offer protection against both predators and extreme heat. These birds, insulated to cope with cold, wet conditions, often find it difficult to keep cool.

During the nesting season, adults obtain small fish and squid from the shallow waters close to the colony. At the end of the season, the birds move away, following northward-flowing ocean currents up the coast. In the east, they may travel 600 miles (1,000 km), to southern Brazil.

Chilean breeders move with the Peru Current, which supports vast stocks of small fish. Groups of penguins may work together or cooperate with sea lions and other predators to concentrate the fish, making them easier to catch.

Magellanic Penguin

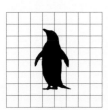

ADÉLIE PENGUIN
Pygoscelis adeliae
Weight
8 lb 13 oz–14 lb 5 oz / 4–6.5 kg
Journey length
600–2,175 miles /
1,000–3,500 km

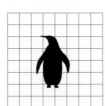

MAGELLANIC PENGUIN
Spheniscus magellanicus
Weight
10–15 lb 7 oz / 4.5–7 kg
Journey length
60–600 miles /
100–1,000 km

EMPEROR PENGUIN
Aptenodytes forsteri
Weight
42–101 lb / 19–46 kg
Journey length
600–1,500 miles /
1,000–2,500 km

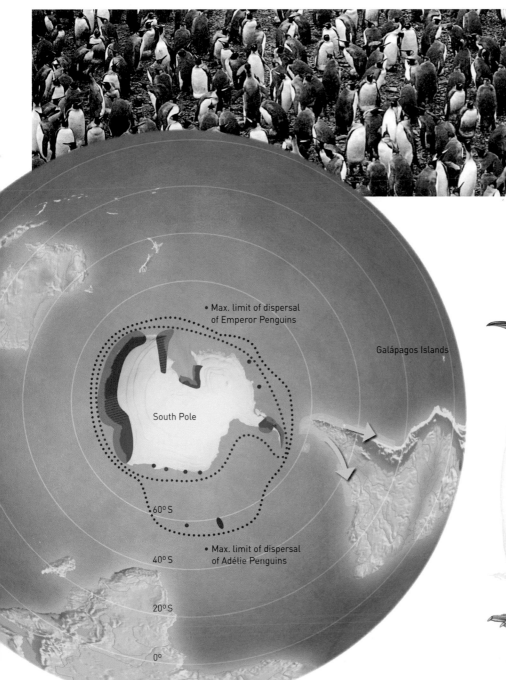

• Max. limit of dispersal
of Emperor Penguins

Galápagos Islands

South Pole

60° S

• Max. limit of dispersal
of Adélie Penguins

40° S

20° S

0°

Emperor Penguin

BY THE TIME A FEMALE EMPEROR PENGUIN returns to the breeding grounds, the male—which arrives first at the start of the breeding season (above)—may not have fed for up to 90 days and has lost some 45 percent of his body weight. He then faces a trek over the ice of up to 125 miles (200 km) to reach the open sea.

If the female arrives after the egg has hatched, the male can feed the chick for up to two weeks with a milky secretion from his esophagus.

Albatrosses

These solitary travelers return to the same island and the same mate each year.

The largest species of albatrosses have the longest wingspans of any flying birds, exceeding 11 feet (3.5 m) in the Wandering Albatross. All are consummate fliers, gliding on outstretched wings in and out of the wave troughs, where they find enough uplift from the wind to maintain their soaring flight for days at a time.

Because they rely almost entirely on soaring—indeed they find flapping flight extremely tiring—albatrosses live only in areas of near-constant wind. In calm conditions they are forced to settle on the sea and wait for the wind to pick up again. The southern oceans, where the winds blow almost continuously in the "roaring forties" and "furious fifties," are home to 10 of the 14 albatross species. Nesting on remote oceanic islands, these birds make extraordinary journeys searching for food: some may regularly travel right around the globe in these latitudes. The other four species inhabit the central and northern Pacific, also nesting on small islands. They disperse across the Pacific, but do not leave it on longer journeys.

ALBATROSSES RARELY reach the North Atlantic. Once there, a bird is unlikely to be able to cross the doldrums—an area of notorious calm north of the equator—and is probably trapped for the rest of its life.

In 1967, a Black-browed Albatross was seen in the Firth of Forth, Scotland. It was there in 1968 and 1969. Then, in 1972, what was almost certainly the same bird appeared in the Shetland Islands. It returned most years for the breeding season, and was there again in summer 1994. For several years it built a nest, without finding a mate to share it.

Black-browed Albatross

THE WANDERING ALBATROSS takes almost a full year to complete its breeding cycle, so nests only every second year. The single egg is incubated for 11 weeks. Once it has hatched, the chick is brooded by one of its parents for the first four weeks, then left while the adults travel to find food. Since they return only every few days, the chick grows slowly and is nine months old before it is ready to leave the nest.

Young Wandering Albatrosses take time to reach maturity and do not make a first breeding attempt until they are at least nine years old. This slow rate of reproduction is high enough to maintain numbers of this long-lived species. There are authentic records of birds aged 40 years or more.

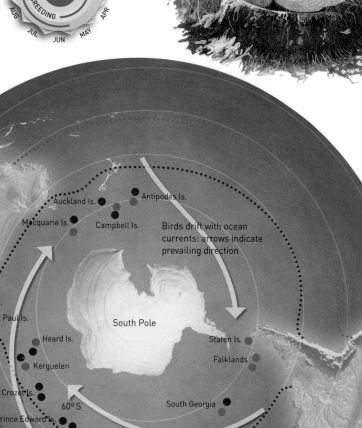

Wandering Albatross

Birds drift with ocean currents: arrows indicate prevailing direction

Auckland Is.
Antipodes Is.
Macquarie Is.
Campbell Is.
Amsterdam Is.
St Paul Is.
South Pole
Heard Is.
Staten Is.
Falklands
Kerguelen
Crozet Is.
South Georgia
60° S
Prince Edward Is.
Marion Is.
Gough Is.
40° S
Tristan da Cunha
20° S
0°

•••••• Max. limit of dispersal of Wandering Albatrosses

•••••• Max. limit of dispersal of Black-browed Albatrosses

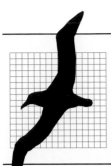

WANDERING ALBATROSS
Diomedea exulans
Wingspan
8 ft 2 in–11 ft 6 in / 2.5–3.5 m
Weight 13 lb 12 oz–25 lb /
6.25–11.3 kg
Journey length
3,000–12,500 miles /
5,000–20,000 km

BLACK-BROWED ALBATROSS
Diomedea melanoptris
Wingspan
7 ft 6 in–8 ft 6 in / 2.3–2.6 m
Weight
6 lb 10 oz–11 lb / 3–5 kg
Journey length
3,000–9,300 miles /
5,000–15,000 km

WAVED ALBATROSS
Diomedea irrorata
Wingspan
7 ft 10 in / 2.4 m
Weight
46 lb / 21 kg
Journey length
950 miles / 1,500 km

Waved Albatross

THE WAVED ALBATROSS is the only species confined to the tropics. About 12,000 breeding pairs nest on Española in the Galápagos, and a colony of 50 pairs was discovered on La Plata, off the Ecuador coast, in 1974.

Waved Albatrosses are unusual in not building a nest but laying their egg on the bare ground. During incubation, the adult may shuffle around, moving the egg up to 130 feet (40 m) from where it was laid.

The main food of most albatrosses is squid, which live in deep water during the day but move to the surface in great schools at night. Albatrosses are normally solitary feeders, but several different species may gather at a rich food source. Humans have long provided food for them: generations ago the debris from whaling ships, today the galley waste from ocean-going cruise ships

Humans equally pose one of the major threats to albatrosses— the birds become trapped in near-invisible monofilament nets or hooked on fishing lines. Longline fishing is now considered the most serious global threat to albatrosses, with perhaps 100,000 being killed every year.

Shearwaters and petrels

Ten-week-old shearwater chicks undertake 6,000-mile (2,000-km) migrations.

A group of about 90 species of highly specialized sea birds that spend by far the greater part of their lives at sea, the shearwaters and petrels come in to land solely to breed. Since many of the smaller species are vulnerable to predators when they are breeding, they nest in burrows that they visit only at night.

All of these birds have a well-developed sense of smell that is used in two ways. First, it enables them to locate sources of food—largely the floating carcasses of dead fish and marine mammals, and the offal and other waste products jettisoned overboard from fishing vessels—over great distances in the wide expanses of the oceans, even at night. Second, it helps in pinpointing the nesting burrow, which the incoming birds have to find among a colony of

thousands of pairs and in darkness. These birds are well known for the persistent, musky odor that clings to their plumage, and it is thought that this may be distinctive enough to allow an individual to recognize its own nest site.

Great Shearwaters breed in the South Atlantic, on the Falkland Islands and on islands in the Tristan da Cunha group, where there are between six and nine million pairs. Outside the breeding season, Greater Shearwaters move around the Atlantic Ocean in a giant loop. They leave the colonies in March and April, heading northwest across the equator to the rich fishing waters off eastern North America, which they reach in May. They spend the whole summer here, then gradually move east across the North

ONE OF THE WORLD'S most abundant birds, the Wilson's Storm-Petrel breeds on almost every subantarctic island and on every suitable stretch of coast in Antarctica. Several colonies in excess of half a million pairs are known.

These birds feed on krill and may be threatened by the new industry in fishing for these planktonic crustaceans. Human intervention has also been of benefit to storm-petrels, however. The virtual elimination of great whales from the southern oceans has removed a major competitor.

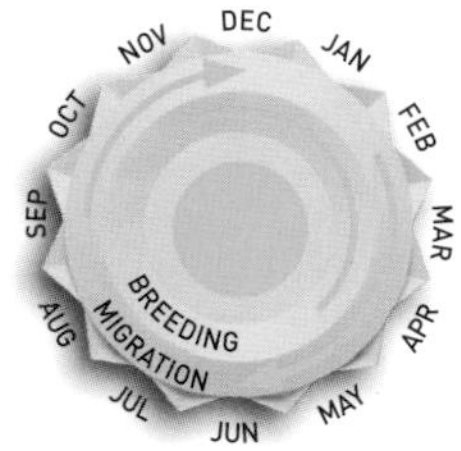

Wilson's Storm-Petrel

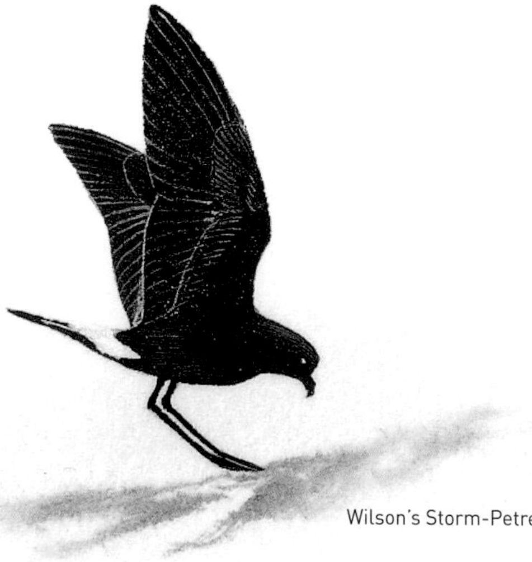

Wilson's Storm-Petrel

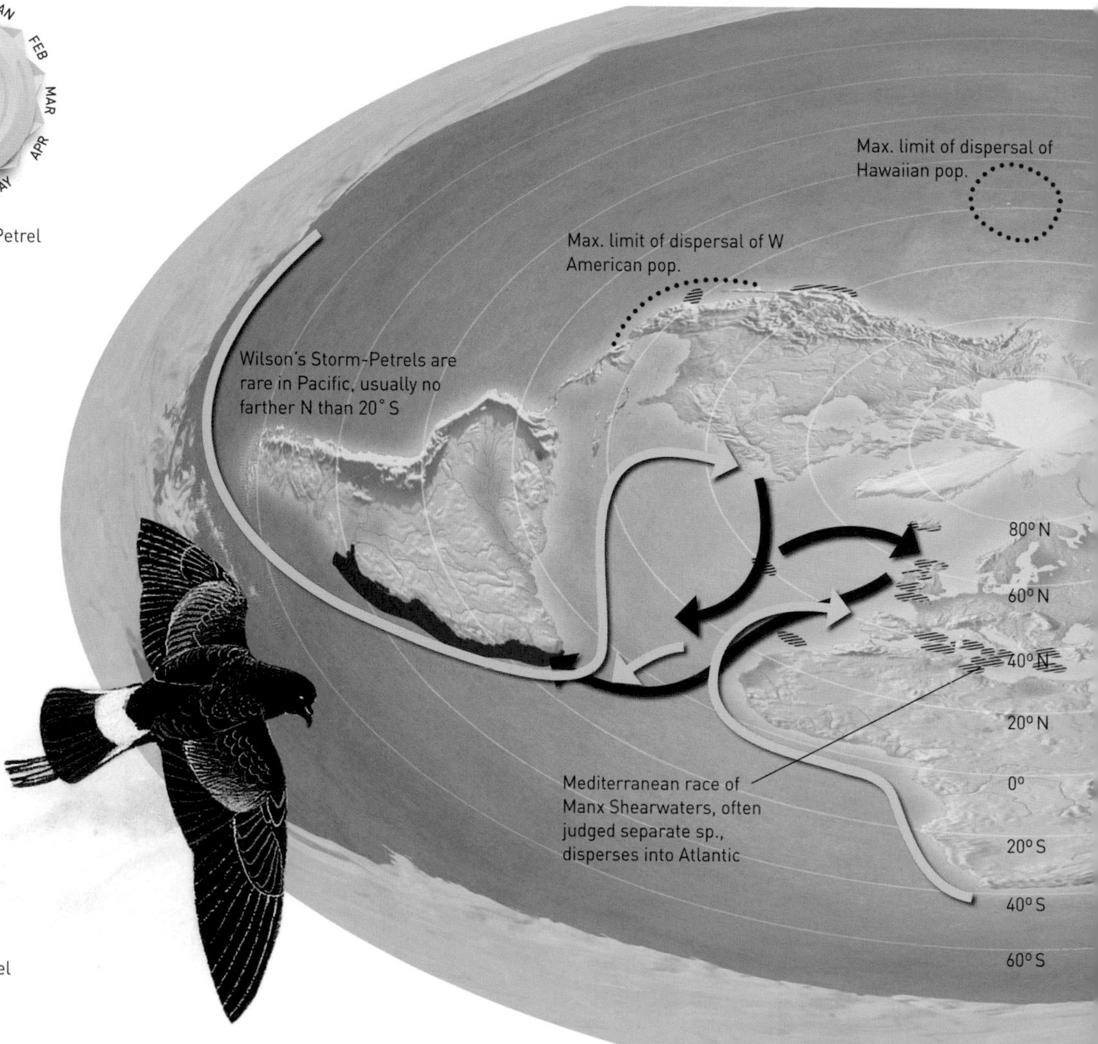

Max. limit of dispersal of Hawaiian pop.

Max. limit of dispersal of W American pop.

Wilson's Storm-Petrels are rare in Pacific, usually no farther N than 20° S

Mediterranean race of Manx Shearwaters, often judged separate sp., disperses into Atlantic

80° N
60° N
40° N
20° N
0°
20° S
40° S
60° S

WILSON'S STORM-PETREL
Oceanites oceanicus
Wingspan
15–17 in / 38–42 cm
Weight
1¼–1½ oz / 34–45 g
Journey length
3,000–9,300 miles /
5,000–15,000 km

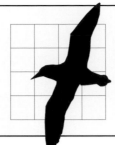

MANX SHEARWATER
Puffinus puffinus
Wingspan
30–35 in / 76–89 cm
Weight
12 oz–1 lb 4 oz / 350–575 g
Journey length
5,300–7,800 miles /
8,500–12,500 km

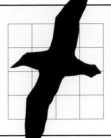

SHORT-TAILED SHEARWATER
Puffinus tenuirostris
Wingspan
3 ft–3 ft 3 in / 95–100 cm
Weight
1 lb 1 oz–1 lb 12 oz / 480–800 g
Journey length
6,800–8,500 miles /
11,000–13,500 km

Manx Shearwater
(European pop.)

Max. limit of dispersal of
New Zealand pop.

MANX SHEARWATERS breed on islands in the North Atlantic. The major colonies are off the west coasts of Britain, Ireland, and France, with the largest, 150,000 pairs—half the world population—on Rhum, off western Scotland.

Manx Shearwater chicks fledge at 10 weeks and set off immediately on migration, crossing the equator to winter off the coasts of Brazil and Argentina. One chick, banded shortly before it fledged on Skokholm Island, Wales, was found off Brazil only 17 days later, having traveled almost 6,000 miles (10,000 km).

Atlantic. They are abundant off the European continental shelf in July and August, by which time they are heading due south again. Some birds are back at their breeding colonies by the end of August; others make a wide, looping sweep in the South Atlantic, completing a figure eight before arriving back at the breeding grounds in mid-September.

Storm-petrels may seem too small and fragile to survive in the stormy oceans where they live, but they are well adapted to their marine existence, feeding on the wing by fluttering close to the sea's surface and scooping up krill or plankton. They also patter with their feet on the water, which may have the effect of attracting their prey.

A RICH RESOURCE

Short-tailed Shearwater

Human exploitation of birds has often been responsible for their near extinction, but where matters are regulated and monitored, populations can survive a harvest of eggs or young.

Short-tailed Shearwaters breed on islands off the coasts of Australia and Tasmania. There are perhaps 23 million birds worldwide, half of which nest in Tasmania. Here, about 600,000 so-called mutton bird chicks are culled annually. The population is monitored to make sure that it withstands this level of harvest.

Nothing is wasted: the feathers are used in upholstery and the down to fill sleeping bags; the meat is eaten; the fat is added to cattle feed; and the oil is used by the pharmaceutical industry.

Skuas, jaegers, gulls, and auks

One group of migrants survives by pirating food from other species.

Gulls and skuas/jaegers are closely related; indeed, the six species of skuas and jaegers clearly evolved from gulls. All skuas and jaegers are great travelers. The four species breeding in the northern hemisphere, Great, Arctic, Long-tailed, and Pomarine, all nest in northerly latitudes. Of these, the great majority of Arctic and Long-tailed jaegers—and at least some of the other two species—winter in the oceans of the southern hemisphere. The South Polar Skua—one of the two species nesting in the southern hemisphere—is known to visit the North Atlantic regularly during the southern winter. A South Polar Skua, banded as a chick on an island in Antarctica at 65°S, was shot six months later in western Greenland at 65°N, a distance of more than 9,300 miles (15,000 km), one of the longest journeys ever recorded for a banded bird.

The gulls are among the most widespread and successful

Long-tailed Jaeger

LIKE OTHER SKUAS, the Long-tailed Jaeger is a predator and a pirate, pursuing other birds until they disgorge or drop the food they are carrying.

In the breeding season, jaegers may nest close to colonies of other sea birds and wait for the adults bringing food to their chicks. Once a jaeger has begun to chase its target, few birds are able to get away.

Migrant jaegers harry victims from above, forcing them down toward the sea. Slowed by the weight of food in its beak or crop, the victim usually drops it. The jaeger then swoops down to take the food for itself.

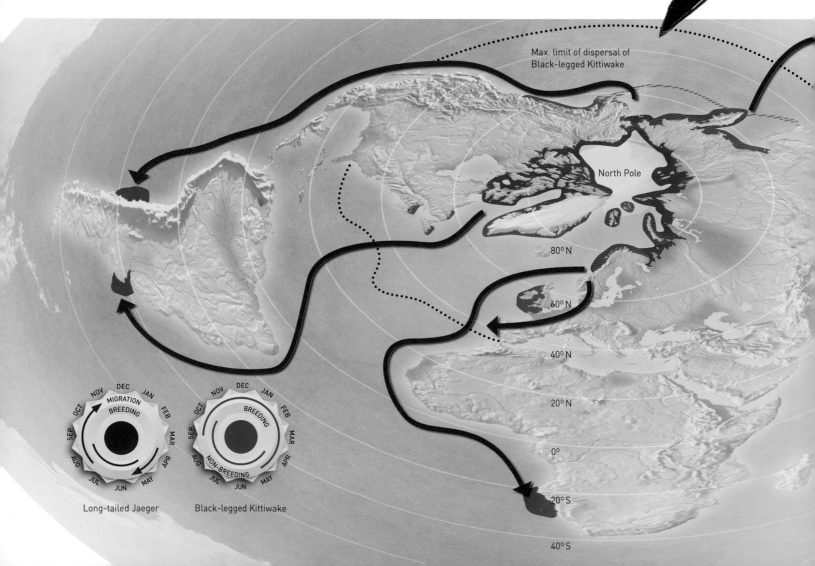

Max. limit of dispersal of
Black-legged Kittiwake

North Pole

80° N

60° N

40° N

20° N

0°

20° S

40° S

Long-tailed Jaeger

Black-legged Kittiwake

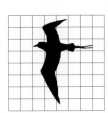

LONG-TAILED JAEGER
Stercorarius longicaudus
Wingspan
3 ft 5 in–3 ft 10 in/
1.05–1.17 m
Weight 9–13 oz/250–360 g
Journey length
5,000–9,000 miles/
8,000–14,400 km

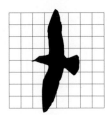

BLACK-LEGGED KITTIWAKE
Rissa tridactyla
Wingspan 3 ft 1 in–3 ft 11 in/
95–120 cm
Weight 11 oz–1 lb 3 oz/
300–535 g
Journey length
300–3,000 miles/
500–5,000 km

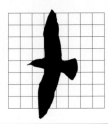

IVORY GULL
Pagophila eburnea
Wingspan 3 ft 6 in–3 ft 11 in/
1.08–1.2 m
Weight 1 lb 2 oz–1 lb 9 oz/
500–700 g
Journey length
125–1,250 miles/
200–2,000 km

sea birds. There are some 45 species, which occur in both hemispheres, although they are more common in the north. These are gregarious birds, breeding in colonies on coastal and some inland marshes and feeding throughout the year in large flocks. Colonies may be tens of thousands strong, particularly if there is a major source of food nearby. Following an elaborate courtship ritual of display and calling, gulls pair for life. A pair usually returns to the same nest site each year, often the colony where one, but not both, of the birds was hatched.

Outside the breeding season most gulls undertake migratory or dispersive movements, or both, sometimes wintering well out to sea. Some are long-distance travelers, although others of the same species may, in a different part of the range, be fairly sedentary. Black-headed Gulls breeding in Japan, for example, migrate south almost to the equator and winter on the islands of Indonesia. Those that breed in Britain, by contrast, simply disperse over comparatively short distances, usually westward to avoid cold weather. The most traveled reach only as far as Spain.

All of the 22 species of auks are confined to the northern hemisphere, where they breed on the coasts, mainly on cliff ledges or in burrows, and winter at sea. Many species extend into the high Arctic, where they are among the most common birds, nesting in huge colonies and feeding on the fish and plankton that abound in the northern seas.

Auks are not good fliers. Although—unlike penguins—they have retained their powers of flight, their short, stubby wings are better suited to underwater swimming. Parents and chicks setting off from the breeding areas on a southerly migration toward the wintering grounds normally start by swimming. The Thick-billed Murres that breed in Svalbard winter off southern Greenland some 1,900 miles (3,000 km) away, and probably cover much of that distance in the water.

THE BLACK-LEGGED KITTIWAKE (below) breeds on the coasts of the North Atlantic, into the Arctic. Outside the breeding season its movements are more dispersive than truly migratory.

Many adults stay within 250–300 miles (400–500 km) of their colony, but the juveniles gradually move farther and farther away from their birthplace in their first two or three winters, reaching their maximum distance (up to 2,175 miles/3,500 km) after three years. They return to the colony to breed when four or five years old.

Black-legged Kittiwake

IVORY GULLS (above) are among the world's most northerly breeding birds, nesting in small colonies on Franz Josef Land, Svalbard, Greenland, and the islands of Arctic Canada. Some nests are on cliff ledges, others have been found on "nunataks," rocky peaks projecting from inland ice sheets. One colony made its home on a floating "ice island" some 1,600 square feet (150 m²) in area, carved from a glacier. The island was about 25 miles (40 km) from the nearest land, and the 75 to 100 pairs of Ivory Gulls were nesting among the rocks and stones that littered its surface.

Terns

Arctic Terns are the world's migratory champions, circumnavigating the globe every year.

Of all birds, the Arctic Tern performs the longest known regular migration. These birds breed in the northern hemisphere, from temperate latitudes to the most northerly land in the world: the tip of Greenland, all the islands of Arctic Canada and Siberia, and Svalbard. They winter on the edge of the Antarctic pack ice. A straight-line distance from the Arctic to the Antarctic is about 9,300 miles (15,000 km). Few birds, however, travel in straight lines, and the Arctic Tern is no exception.

Terns nesting in the Arctic areas of eastern Canada and Greenland head southeast toward Europe to join the European, Svalbard, and Russian Arctic breeders. They all then head south down the coast of Iberia and western Africa. Some split off around Cape Verde and make their way down the South American coast, but the majority stay off the coast of Africa. Once past the Cape of Good Hope, they continue to head south for Antarctica, but are gradually pushed eastward by the prevailing winds until they are in the southern Indian Ocean.

As the pack ice retreats during the southern summer, the terns follow it southward, since its boundary usually marks an area of good feeding. By continuing south, the birds enter a zone of westerly winds and, when it is time to head back north, are thus south of the Atlantic Ocean again. They follow more or less the same route back up the coast of Africa. Once across the equator, they fan out, with the birds from Canada taking a more northwesterly route, while European breeders hug the coast.

Arctic Terns feed mainly on small fish. One of their most important prey species in northern waters is the sand eel, a preference that has posed problems for some Arctic Terns, notably those breeding in Shetland, Scotland. Here, there were about 40,000 pairs in 1980, but after a series of complete breeding failures, numbers had halved by 1990. This has been linked to scarcity of sand eels, which have been commercially fished with no limits on numbers taken.

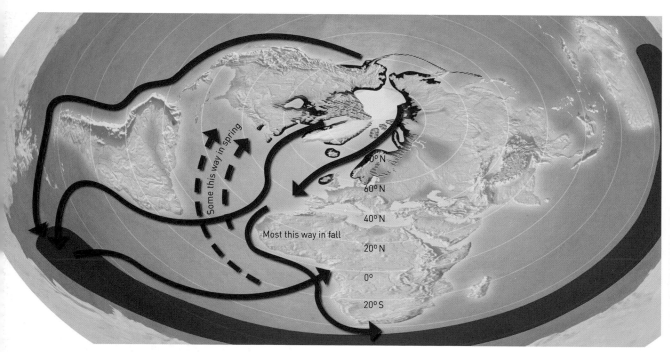

Some this way in spring

Most this way in fall

80° N
60° N
40° N
20° N
0°
20° S

THE COMPLETE MIGRATORY round trip for a single Arctic Tern (right) could be as much as 18,650–25,000 miles (30,000–40,000 km), almost the same distance as the circumference of the globe. Since the oldest banded Arctic Tern was 26 years old, this amounts to more than 620,000 miles (1 million km) in the course of a lifetime.

Another feature of the Arctic Tern's phenomenal travels is that those birds breeding on the islands north of the Arctic Circle experience the 24 hours of daylight that occur there in the summer months, and then, by making the long flight to south of the Antarctic Circle, find the same conditions during the southern summer. In this way, they enjoy more hours of daylight in a year than any other living thing.

Arctic Tern

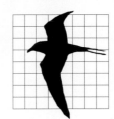

ARCTIC TERN
Sterna paraisaea
Wingspan
29–33 in / 75–85 cm
Weight
3–4 oz / 80–120 g
Journey length
9,300–12,500 miles /
15,000–20,000 km

The Shetland sand eel fishery was closed by government order in March 1991, ostensibly to conserve stocks for future fishing. It is nevertheless hoped that as sand eel stocks recover, the terns—and many other sea bird species that have suffered food shortages in this area—will benefit. Early indications are that this is working.

These birds are colonial nesters. In the high Arctic, colonies are generally small, consisting of a few to perhaps 100 pairs, but farther south, colonies are much larger, with some holding as many as 10,000 pairs. Nests are usually placed only 10–13 feet (3–4 m) apart. The clutch of two or three eggs is laid in a shallow scrape in the sand or pebbles, and newly hatched chicks are covered in a camouflaging gray and buff down that makes them difficult to see when they are crouched on the ground.

Arctic Terns are extremely aggressive in defense of their nests and make strenuous and concerted efforts to drive any intruder away from the area of the colony. Foxes intending predation and humans wanting to count the number of nests receive the same treatment. The adult birds from all over the colony rise up and make repeated swoops at the head, often striking with the bill, which is capable of drawing blood from an unprotected scalp.

Almost migrations

Conventional migrations are regular, taking place at the same time each year, and fixed in distance and direction. Others are less predictable. In some falls few, if any, birds leave the breeding area, while in others the whole population may shift thousands of miles. Such movements are known as irruptions, since they are most noticeable in the areas to which the birds move en masse. Here they "irrupt," often in spectacular numbers.

Irruptive species breed largely in the north and divide into three broad categories: the raptors and owls, the seed- and berry-eaters, and the birds of the steppes, such as Pallas' Sandgrouse. All have one thing in common, a relatively restricted diet that is itself subject to wide fluctuations in abundance. Thus, irruptive raptors and owls feed on voles and lemmings, which have a four- to five-year cycle of scarcity and abundance. Similarly, most irrupting seed- and berry-eaters feed on trees such as spruce, pine, and birch, which after a heavy crop may rest the following year. Sandgrouse take a variety of seeds, but live in areas subject to droughts that prevent the plants from flowering and seeding.

Reliant on voles and lemmings to feed themselves and their young, Common Buzzards are forced south when rodent populations crash.

Irruptions

Failures of prey species and of seed or berry crops force reluctant migrants south.

The most commonly irruptive birds of prey include the Northern Goshawk and Rough-legged Hawk. Among the owls, the Snowy, Northern Hawk, and Great Horned irrupt most frequently. All five species breed in the forests and tundra of the far north, the Great Horned Owl in North America only, the other species in both North America and Eurasia. They are equally irruptive in both continents.

These species feed on prey that have marked cycles of abundance. The reason why prey species show such strong cyclical fluctuations in numbers is uncertain, but may be due to variations in the climate combined with the availability of their own food supply, largely seeds and plants. Whatever the cause, the effect on predators of changes in their prey populations is marked.

In some regions, the Northern Goshawk and Great Horned Owl depend heavily on grouse and hares. In North America, the snowshoe hare, which has an approximately 10-year cycle of abundance, is the preferred target. In Eurasia, the principal food is grouse, which peaks and troughs about every four years. Thus, in North America, the Northern Goshawk and Great Horned Owl irrupt every 10 years, while in Eurasia the Goshawks move every four years.

Rough-legged Hawks and Snowy and Northern Hawk owls all feed on small rodents, mainly voles and lemmings, which also fluctuate in approximately four-year cycles. In years when voles and lemmings abound, hawks and owls breed very successfully, rearing large broods of young. Since food is plentiful, many of these birds survive the winter in the same area, so bird numbers are higher than usual in the following spring. Once the crash in vole and lemming numbers begins, the predators, having reached a high level of population themselves, either have to move or risk starvation. It is at this point that the predators migrate to find alternative food supplies.

Pallas' Sandgrouse breed in central Asia, Russia, Mongolia, and northern China, and normally make short-distance movements to the south. At infrequent intervals, large numbers move west into Europe, reaching as far as Britain and Ireland. The weather seems to play the major part in stimulating such movements: drought probably results in a poor seed crop, then exceptional cold in the following spring means that nest sites remain snow-covered, forcing the birds to move away.

More numerous than irruptive predators are seed- and berry-eaters. Several species of finches breeding in the northern forests

Snowy Owl

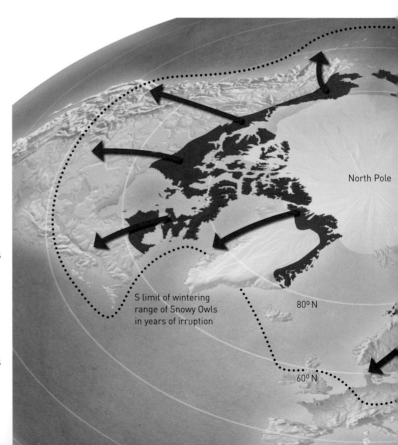

North Pole

S limit of wintering range of Snowy Owls in years of irruption

80° N

60° N

SNOWY OWL
Nyctea scandiaca
Wingspan
4 ft 8 in–5 ft 5 in / 1.42–1.66 m
Weight 2 lb 10 oz–6 lb 8 oz /
1.2–2.95 kg
Journey length
300–3,000 miles /
500–5,000 km

ROUGH-LEGGED HAWK
Buteo lagopus
Wingspan
3 ft 11 in–4 ft 11 in / 1.2–1.5 m
Weight 1 lb 5 oz–3 lb 11 oz /
600 g–1.66 kg
Journey length
300–2,800 miles /
500–4,500 km

RED CROSSBILL
Loxia curvirostra
Wingspan
10–11 in / 25–28 cm
Weight
1¼–1½ oz / 38–42 g
Journey length
300–2,175 miles /
500–3,500 km

North American race
sancti-johannis

Rough-legged Hawk

Eurasian race *lagopus*

WHEN LEMMING NUMBERS are high, Snowy Owls breed well and rear large broods of young. When prey numbers are low, the owls may not be able to breed at all and leave northern areas early on much longer migrations than usual. They head south and stop only when they find a source of food. In "normal" years, Snowy Owls stay in Canada and northern Europe, but in years of irruption, North American breeders may travel to the Gulf States, while Eurasian birds may reach the Mediterranean.

Birds that wander long distances from the breeding grounds are probably unable to return the following spring.

make unusually long mass movements when the seed crop of certain trees fails. Fruiting and seeding trees do not seem to have regular cycles of success and failure, but it is common for a productive year to be followed by one in which the trees appear to rest.

Good years enable northerly breeding tits, finches, and thrushes to survive through that fall and winter, so there is a large breeding population the following spring. These birds, and the young that they rear, then find there are few berries and seeds to eat that fall and are forced to migrate.

ROUGH-LEGGED HAWKS breed on the Arctic tundra of North America and Eurasia. The extent of their migration is dictated by food stocks. When lemmings and voles are plentiful, the birds winter in the northern U.S. and northern Europe. In irruptive years, they may reach Florida and the Gulf coast, western Europe, northern Africa, and the Middle East.

These birds are doubly adapted to fluctuating food supplies in that they can also shift breeding areas, often breeding hundreds of miles south, east, or west of their normal range when small mammals are scarce.

WHEN NORTHERN-LATITUDE trees fail to produce a seed crop, all of the birds that rely on the crop are forced to irrupt. Thus, in any autumn, if one species is irrupting, the chances are that several others will be doing so too.

The graph shows this synchrony by illustrating the numbers of four species of finches found during the annual Christmas count in Chesapeake Bay, Maryland, over 10 winters. Although numbers of individuals vary from bird to bird and year to year, they peak and trough in the same years for all species.

Evening Grosbeak

Red Crossbill

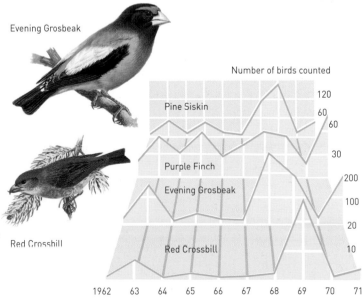

Number of birds counted

Pine Siskin

Purple Finch

Evening Grosbeak

Red Crossbill

1962   63   64   65   66   67   68   69   70   71

ABBREVIATIONS

N	North/northern	**Wn**	Wingspan
S	South/southern	**Wt**	Weight
E	East/eastern	**C** = Central/central	
W	West/western	**NZ** = New Zealand	
		US = United States	

Catalog of migrants

This list follows the order of the chapters in the book. Within each section, birds are listed according to the systematic order.

NORTH AMERICAN MIGRANTS

WESTERN GREBE
Aechmophorus occidentalis
Wn 38–42 in
Wt 2 lb 3 oz–3 lb 12 oz
Breeds WC N America S to Mexico, Apr to Sep
Winters Coasts of California & Mexico, Oct to Mar
Journey length 1,100–2,500 miles

EARED GREBE *Podiceps nigricollis*
Wn 22–23 in **Wt** 9–14 oz
Breeds WC US & S Canada; also Eurasia, May to Sep
Winters California, Gulf of Mexico S to Guatemala, Oct to Apr
Journey length 300–3,000 miles

DOUBLE-CRESTED CORMORANT
Phalacrocorax auritus
Wn 49–54 in
Wt 3 lb 8 oz–4 lb 10 oz
Breeds C & N US, S & E Canada, Apr to Oct
Winters W & S coasts US, Mexico & Cuba, Nov to Mar
Journey length 300–3,400 miles

AMERICAN BITTERN
Botaurus lentiginosus
Wn 41–49 in **Wt** 13 oz–1 lb 4 oz
Breeds US & S Canada, Apr to Oct
Winters S US, Mexico & C America S to Panama, Nov to Mar
Journey length 500–4,200 miles

LEAST BITTERN *Ixobrychus exilis*
Wn 15–23 in **Wt** 4–6 oz
Breeds SE Canada, E & W US, C America, N S America, Mar to Oct
Winters Mexico to N S America, Nov to Feb
Journey length 600–2,500 miles

WOOD DUCK *Aix sponsa*
Wn 26–29 in
Wt 1 lb 2 oz–1 lb 15 oz
Breeds C & S US, Feb to Aug
Winters S US, Mexico & Cuba, Sep to Jan
Journey length 125–1,250 miles

AMERICAN BLACK DUCK
Anas rubripes
Wn 33–37 in **Wt** 2 lb 9 oz–3 lb
Breeds NE US & E Canada, Mar to Sep
Winters SE US, Oct to Feb
Journey length 185–2,175 miles

CANVASBACK *Aythya valisneria*
Wn 32–36 in
Wt 1 lb 14 oz–3 lb 8 oz
Breeds NW US & W Canada, May to Sep
Winters S US & Mexico, Oct to Apr
Journey length 300–3,400 miles

LESSER SCAUP *Aythya affinis*
Wn 27–31 in
Wt 1 lb 12 oz–1 lb 14 oz
Breeds W US & W Canada, May to Sep
Winters S US, Mexico & C America S to Panama, Oct to Apr
Journey length 600–5,600 miles

BUFFLEHEAD *Bucephala albeola*
Wn 21–24 in **Wt** 12–16 oz
Breeds Alaska, W & S Canada, Apr to Aug
Winters E & W coasts of N America, inland S US & N Mexico, Sep to Mar
Journey length 300–3,400 miles

RUDDY DUCK *Oxyura jamaicensis*
Wn 20–24 in **Wt** 14 oz–1 lb 10 oz
Breeds N & W US, SW Canada; also Eurasia, Mar to Sep
Winters S & SE US, Mexico, Oct to Feb
Journey length 600–2,500 miles

BLACK VULTURE *Coragyps atratus*
Wn 55–59 in
Wt 3 lb 5 oz–5 lb 8 oz
Breeds S & SE US, Mar to Sep
Winters S & SE US, C & S America, Oct to Feb
Journey length 60–600 miles

BALD EAGLE
Haliaeetus leucocephalus
Wn 70–90 in
Wt 6 lb 10 oz–15 lb 7 oz
Breeds N US & Canada, Mar to Sep
Winters Alaska to Newfoundland S to S US, Oct to Feb
Journey length 60–2,800 miles

NORTHERN HARRIER
Circus cyaneus
Wn 39–47 in **Wt** 11 oz–1 lb 9 oz
Breeds C US N to N Canada; also Eurasia, Apr to Aug
Winters S US S to W Indies & Colombia, Sep to Mar
Journey length 300–3,000 miles

SHARP-SHINNED HAWK
Accipiter striatus
Wn 20–28 in **Wt** 4–13 oz
Breeds N US, Alaska, S & NW Canada, sedentary in W Indies & S America, Apr to Sep
Winters S US & Mexico, Oct to Mar
Journey length 60–3,000 miles

COOPER'S HAWK *Accipiter cooperii*
Wn 29–37 in
Wt 7 oz–1 lb 2 oz
Breeds S Canada, US & N Mexico, Apr to Sep
Winters US, C America & W Indies, Oct to Mar
Journey length 0–1,250 miles

BROAD-WINGED HAWK
Buteo platypterus
Wn 31–36 in
Wt 7 oz–1 lb 3 oz
Breeds S Canada & E US, Apr to Sep
Winters S Florida to Brazil, Bolivia & Peru, Oct to Mar
Journey length 1,000–8,700 miles

RED-TAILED HAWK
Buteo jamaicensis
Wn 47–53 in
Wt 1 lb 2 oz–2 lb 10 oz
Breeds S Canada & US S to Panama & W Indies, Mar to Sep
Winters US, except extreme N, S to Panama & W Indies, Oct to Feb
Journey length 0–1,850 miles

FERRUGINOUS HAWK
Buteo regalis
Wn 51–55 in **Wt** 2 lb–3 lb 15 oz
Breeds W Canada & NW US, Mar to Sep
Winters SW US & N Mexico, Oct to Feb
Journey length 125–2,500 miles

PRAIRIE FALCON *Falco mexicanus*
Wn 35–43 in
Wt 1 lb 5 oz–2 lb 14 oz
Breeds SW Canada & W US, Apr to Sep
Winters W US to Mexico, Oct to Mar
Journey length 0–1,250 miles

PEREGRINE FALCON
Falco peregrinus
Wn 35–43 in
Wt 1 lb 5 oz–2 lb 14 oz
Breeds Alaska, N Canada & WC US; also Eurasia, May to Sep
Winters E coast of US, coasts of Mexico & W Indies, Oct to Apr
Journey length 1,250–3,750 miles

SORA *Porzana carolina*
Wn 13–15 in **Wt** 2–4 oz
Breeds S Canada, N & C US, Apr to Sep
Winters S US, C America & S America S to N Peru & Guyana, Oct to Mar
Journey length 1,850–4,350 miles

AMERICAN AVOCET
Recurvirostra americana
Wn 30–33 in **Wt** 10 oz–1 lb 1 oz
Breeds S Canada, W & extreme E US, May to Sep
Winters S US to Guatemala, Oct to Apr
Journey length 300–2,800 miles

SEMIPALMATED PLOVER
Charadrius semipalmatus
Wn 16–20 in **Wt** 1–2 oz
Breeds Alaska & N Canada S to Newfoundland, May to Aug
Winters E & W coasts S US to S America, Oct to Mar
Journey length 4,000–8,000 miles

MOUNTAIN PLOVER
Charadrius montanus
Wn 19–23 in **Wt** 2–3 oz
Breeds Montana to New Mexico, Apr to Aug
Winters S California & N Mexico, Sep to Mar
Journey length 370–1,500 miles

AMERICAN GOLDEN PLOVER
Pluvialis dominica
Wn 23–28 in **Wt** 4–7 oz
Breeds Alaska & Arctic Canada, May to Aug
Winters Bolivia & S Brazil to C Argentina, Oct to Mar
Journey length 5,900–8,200 miles

MARBLED GODWIT
Limosa fedoa
Wn 29–33 in **Wt** 9 oz–1 lb 5 oz
Breeds CS Canada & CN US, Apr to Aug
Winters E & W coasts of US S to Chile, Oct to Mar
Journey length 900–5,600 miles

LONG-BILLED CURLEW
Numenius americanus
Wn 31–39 in
Wt 1 lb 9 oz–3 lb 1 oz
Breeds SW Canada & NW US, Apr to Sep
Winters California & Gulf S to Guatemala, Oct to Mar
Journey length 300–3,400 miles

UPLAND SANDPIPER
Bartramia longicauda
Wn 25–26 in **Wt** 4–6 oz
Breeds Alaska, W & S Canada, C & N US, May to Sep
Winters Brazil, Paraguay, Uruguay & Argentina, Oct to Apr
Journey length 5,900–7,500 miles

GREATER YELLOWLEGS
Tringa melanoleuca
Wn 27–29 in **Wt** 4–8 oz
Breeds S Alaska & Canada from British Columbia to Newfoundland, Apr to Aug
Winters S coasts of US S to S America, Oct to Mar
Journey length 1,500–8,000 miles

LESSER YELLOWLEGS
Tringa flavipes
Wn 23–25 in **Wt** 2–4 oz
Breeds E Alaska & Canada E to Hudson Bay, May to Aug
Winters S US, W Indies & S America, Oct to Apr
Journey length 1,500–9,300 miles

SOLITARY SANDPIPER
Tringa solitaria
Wn 21–23 in **Wt** 1–2 oz
Breeds Alaska & S Canada, May to Aug
Winters S US & W Indies S to Peru, N Argentina & Uruguay, Sep to Apr
Journey length 2,175–6,800 miles

SPOTTED SANDPIPER
Actitis macularia
Wn 14–15 in **Wt** 1–2 oz
Breeds Canada & US S to California & S Carolina, May to Sep
Winters Extreme S US, W Indies, C & S America S to Peru, Bolivia & Brazil, Oct to Mar
Journey length 600–6,800 miles

WILSON'S PHALAROPE
Phalaropus tricolor
Wn 15–16 in **Wt** 1–3 oz
Breeds W Canada, W & C US, May to Aug
Winters S America from Bolivia & Peru to Chile & Argentina, Sep to Apr
Journey length 4,475–8,000 miles

SHORT-BILLED DOWITCHER
Limnodromus griseus
Wn 17–20 in **Wt** 3–5 oz
Breeds S Alaska, NC & E Canada, May to Aug
Winters S US, C America, W Indies & S America S to Peru & Brazil, Sep to Mar
Journey length 1,500–5,300 miles

SURFBIRD *Aphriza virgata*
Wn 19–21 in **Wt** 3–5 oz
Breeds Alaska & NW Canada, May to Aug
Winters W coast of N & S America; Galápagos, Sep to Apr
Journey length 300–8,700 miles

BONAPARTE'S GULL
Larus philadelphia
Wn 35–39 in **Wt** 6–7 oz
Breeds Alaska & W Canada, May to Oct
Winters Coastal US S to Mexico & W Indies, Oct to Mar
Journey length 300–3,400 miles

RING-BILLED GULL
Larus delawarensis
Wn 47–61 in
Wt 1 lb 2 oz–1 lb 10 oz
Breeds C & E Canada, CN US, Mar to Aug
Winters N US S to Mexico, Sep to Mar
Journey length 60–2,500 miles

MEW GULL *Larus canus*
Wn 43–51 in **Wt** 11 oz–1 lb 3 oz
Breeds NW N America & N Eurasia; resident in W Europe, May to Sep
Winters Pacific coasts of N America, W Europe, N & Baltic seas & E Asia, Oct to Apr
Journey length 300–4,350 miles

HERRING GULL *Larus argentatus*
Wn 54–61 in
Wt 1 lb 9 oz–3 lb 1 oz
Breeds Alaska, Canada & NE US; also Eurasia, Mar to Oct
Winters NE & NW US S to Mexico, Oct to Mar
Journey length 60–2,500 miles

FORSTER'S TERN *Sterna forsteri*
Wn 28–32 in **Wt** 3–5 oz
Breeds SC Canada, N US, May to Sep
Winters S US S to Guatemala, Sep to Apr
Journey length 125–2,800 miles

LEAST TERN *Sterna antillarum*
Wn 18–21 in **Wt** 1–2 oz
Breeds S US S to C America & W Indies, May to Sep
Winters Coasts of N, C & N S America & W Indies, Sep to Apr
Journey length 60–1,850 miles

BLACK-BILLED CUCKOO
Coccyzus erythropthalmus
Wn 14–16 in **Wt** 1–2 oz
Breeds S Canada & US, except extreme W & S, May to Oct
Winters NW S America to Peru, Sep to Apr
Journey length 2,175–4,700 miles

YELLOW-BILLED CUCKOO
Coccyzus americanus
Wn 15–18 in **Wt** 1–3 oz
Breeds SE Canada & US, except NW, S to C Mexico & W Indies, May to Sep
Winters S America from Venezuela & Colombia to Argentina, Sep to Apr
Journey length 1,250–5,000 miles

LONG-EARED OWL *Asio otus*
Wn 35–39 in **Wt** 7–15 oz
Breeds S Canada & US except extreme S, Mar to Aug
Winters US S to N Mexico, Sep to Feb
Journey length 0–3,000 miles

FLAMMULATED OWL
Otus flammeolus
Wn 13–15 in **Wt** 2–3 oz
Breeds W US S to Mexico, Apr to Sep
Winters Mexico to Guatemala, Sep to Mar
Journey length 300–3,000 miles

BURROWING OWL
Athene cunicularia
Wn 20–24 in **Wt** 5–8 oz
Breeds SC Canada, W & C US, Mexico, C & S America, Apr to Oct
Winters SW US, Mexico, C & S America, Oct to Apr
Journey length 0–1,250 miles

CHUCK-WILL'S-WIDOW
Caprimulgus carolinensis
Wn 26–27 in **Wt** 2–2¼ oz
Breeds SC & E US, Apr to Sep
Winters Extreme SE US, W Indies, N S America, Sep to Mar
Journey length 300–2,500 miles

VAUX'S SWIFT *Chaetura vauxi*
Wn 13–15 in **Wt** ½–1 oz
Breeds S Alaska, SW Canada & W US, May to Sep
Winters Venezuela & Colombia, Oct to Apr
Journey length 600–4,350 miles

BROAD-TAILED HUMMINGBIRD
Selasphorus platycercus
Wn 5–6 in **Wt** 1/14 oz
Breeds W & SW US, N Mexico, Apr to Aug
Winters Mexico to Guatemala, Sep to Mar
Journey length 125–2,800 miles

RUFOUS HUMMINGBIRD
Selasphorus rufus
Wn 5–6 in **Wt** 1/15 oz
Breeds Alaska, SW Canada & NW US, Apr to Aug
Winters Mexico, Sep to Mar
Journey length 300–3,750 miles

BELTED KINGFISHER *Ceryle alcyon*
Wn 18–20 in **Wt** 4–7 oz
Breeds Alaska, C & S Canada & US, Apr to Sep
Winters S US, Mexico & C America S to Panama, Sep to Apr
Journey length 0–4,700 miles

EASTERN KINGBIRD
Tyrannus tyrannus
Wn 11–12 in **Wt** 1½ oz
Breeds S Canada & US except SW, Apr to Sep
Winters S America S to N Argentina, Oct to Mar
Journey length 1,850–6,800 miles

GRAY KINGBIRD
Tyrannus dominicensis
Wn 11–12 in **Wt** 1⅔–1¾ oz
Breeds S Florida & W Indies, Mar to Sep
Winters W Indies to N S America, Oct to Feb
Journey length 125–1,500 miles

WESTERN KINGBIRD
Tyrannus verticalis
Wn 11–12 in **Wt** 1⅔–1¾ oz
Breeds SW Canada, W US & N Mexico, Apr to Sep
Winters C Mexico S to Nicaragua, Oct to Mar
Journey length 1,100–3,400 miles

GREAT CRESTED FLYCATCHER
Myiarchus crinitus
Wn 10–11 in **Wt** 1½ oz
Breeds S & E Canada, E & C US, Apr to Oct
Winters Extreme S Florida S to Mexico & N S America, Oct to Mar
Journey length 155–3,400 miles

OLIVE-SIDED FLYCATCHER
Contopus borealis
Wn 9–11 in **Wt** 1–1½ oz
Breeds S Alaska, W & S Canada, N & W US, Apr to Sep
Winters C & S America S to Peru, Sep to Mar
Journey length 1,850–5,600 miles

EASTERN WOOD-PEWEE
Contopus virens
Wn 9–10 in **Wt** 1½ oz
Breeds S & E Canada & E US, Apr to Sep
Winters Costa Rica, Colombia & Venezuela S to Peru & W Brazil, Oct to Mar
Journey length 1,250–4,475 miles

WESTERN WOOD-PEWEE
Contopus sordidulus
Wn 9–10 in **Wt** 1–1½ oz
Breeds S Alaska, W Canada, W US & N Mexico, Apr to Sep
Winters Panama & Venezuela S to Bolivia, Oct to Mar
Journey length 1,100–4,700 miles

EASTERN PHOEBE *Sayornis phoebe*
Wn 9–11 in **Wt** 1 oz
Breeds C & S Canada, C & E US, Apr to Oct
Winters S & SE US S to S Mexico, Oct to Apr
Journey length 0–3,400 miles

GRAY FLYCATCHER
Empidonax wrightii
Wn 9–10 in **Wt** ½ oz
Breeds W US, Apr to Sep
Winters SW US & Mexico, Oct to Mar
Journey length 185–1,850 miles

DUSKY FLYCATCHER
Empidonax oberholseri
Wn 8–9 in **Wt** ½ oz
Breeds W Canada & W US, Apr to Sep
Winters SW US & Mexico, Oct to Mar
Journey length 185–2,175 miles

HAMMOND'S FLYCATCHER
Empidonax hammondii
Wn 8–9 in **Wt** ½ oz
Breeds S Alaska, W Canada & W US, May to Sep
Winters SW US & Mexico S to Nicaragua, Oct to Apr
Journey length 300–4,000 miles

LEAST FLYCATCHER
Empidonax minimus
Wn 7–8 in **Wt** ½ oz
Breeds S Canada & N US, May to Sep
Winters Mexico to Panama, Oct to Apr
Journey length 1,250–4,475 miles

ACADIAN FLYCATCHER
Empidonax virescens
Wn 8–9 in **Wt** ½ oz
Breeds S Ontario & E US, May to Sep
Winters Costa Rica to Ecuador & Venezuela, Oct to Apr
Journey length 1,100–3,400 miles

WILLOW FLYCATCHER
Empidonax traillii
Wn 8–9 in **Wt** ½ oz
Breeds SW Canada, N & C US, May to Sep
Winters S America S to Argentina, Oct to Apr
Journey length 2,175–6,000 miles

ALDER FLYCATCHER
Empidonax alnorum
Wn 8–9 in **Wt** ½ oz
Breeds Alaska, W & S Canada, NE US, May to Sep
Winters C America S to Panama, Oct to Apr
Journey length 2,800–4,700 miles

YELLOW-BELLIED FLYCATCHER
Empidonax flaviventris
Wn 7–9 in **Wt** ½ oz
Breeds S Canada, NE US, May to Sep
Winters E Mexico S to E Panama, Oct to Apr
Journey length 2,800–4,700 miles

WESTERN FLYCATCHER
Empidonax difficilis
Wn 7–9 in **Wt** ½ oz
Breeds British Columbia & W US, Apr to Oct
Winters N Mexico to Honduras, Oct to Apr
Journey length 300–4,350 miles

VIOLET-GREEN SWALLOW
Tachycineta thalassina
Wn 10–11 in **Wt** ½ oz
Breeds Alaska, W Canada, W US & Mexico, Apr to Oct
Winters Mexico & C America S to Panama, Oct to Mar
Journey length 600–5,600 miles

PURPLE MARTIN *Progne subis*
Wn 13–15 in **Wt** ¾–1 oz
Breeds S Canada, US & Mexico, Apr to Oct
Winters W Indies, S America S to Brazil, Nov to Mar
Journey length 600–5,900 miles

CLIFF SWALLOW
Hirundo pyrrhonota
Wn 12–13 in **Wt** ¾–1 oz
Breeds Alaska, Canada, US & N Mexico, Apr to Oct
Winters S America S to Argentina, Nov to Mar
Journey length 1,250–6,800 miles

HOUSE WREN *Troglodytes aedon*
Wn 5–7 in **Wt** ¼–½ oz
Breeds S Canada, US & N Mexico, Mar to Oct
Winters S US & Mexico, Oct to Mar
Journey length 60–1,500 miles

WINTER WREN
Troglodytes troglodytes
Wn 5–6 in **Wt** ⅕–⅖ oz
Breeds S Alaska, S Canada & N US; also Eurasia, Mar to Sep
Winters S US, Oct to Feb
Journey length 125–1,850 miles

MARSH WREN *Cistothorus palustris*
Wn 5–8 in **Wt** ½ oz
Breeds S Canada & N US, Apr to Oct
Winters S US & Mexico, Oct to Mar
Journey length 185–2,500 miles

SEDGE WREN *Cistothorus platensis*
Wn 5–7 in **Wt** ¼–½ oz
Breeds S & SE Canada, NE US, Apr to Oct
Winters SE US & Mexico, Oct to Mar
Journey length 600–2,500 miles

RUBY-CROWNED KINGLET
Regulus calendula
Wn 5–6 in **Wt** ¼ oz
Breeds Alaska, S Canada & W US, Apr to Oct
Winters S & E US, Mexico & Guatemala, Oct to Mar
Journey length 185–5,900 miles

BLUE-GRAY GNATCATCHER
Polioptila caerulea
Wn 5–7 in **Wt** ¼ oz
Breeds S US, Mexico & Cuba, Mar to Oct
Winters S Mexico S to Honduras, Nov to Feb
Journey length 60–1,250 miles

EASTERN BLUEBIRD *Sialia sialis*
Wn 9–12 in **Wt** 1–1¼ oz
Breeds SE Canada, E US & C America S to Nicaragua, Mar to Oct
Winters SE US S to Nicaragua, Oct to Mar
Journey length 0–2,500 miles

WESTERN BLUEBIRD
Sialia mexicana
Wn 9–12 in **Wt** 1–1¼ oz
Breeds SW Canada, W US & Mexico, Mar to Oct
Winters SW US & Mexico, Oct to Mar
Journey length 0–2,800 miles

MOUNTAIN BLUEBIRD
Sialia currucoides
Wn 10–12 in **Wt** 1–1¼ oz
Breeds SW Canada, W US, Apr to Oct
Winters SW US, N Mexico, Oct to Mar
Journey length 0–3,000 miles

WOOD THRUSH
Hylocichla mustelina
Wn 11–13 in **Wt** 1½–2½ oz
Breeds SE Canada & E US, Mar to Sep
Winters Mexico to Panama, Oct to Mar
Journey length 600–3,750 miles

VEERY *Catharus fuscescens*
Wn 11–12 in **Wt** ¾–1½ oz
Breeds S Canada & N US, May to Sep
Winters N S America, Oct to Apr
Journey length 2,500–5,000 miles

SWAINSON'S THRUSH
Catharus ustulatus
Wn 10–11 in **Wt** ¾–1½ oz
Breeds Alaska, W & S Canada, W US, May to Oct
Winters Mexico to Argentina, Oct to Apr
Journey length 1,500–6,800 miles

HERMIT THRUSH *Catharus guttatus*
Wn 9–11 in **Wt** 1 oz
Breeds S Alaska, S Canada, NE & W US, Apr to Oct
Winters S US & C America S to El Salvador, Oct to Mar
Journey length 125–5,000 miles

NORTHERN SHRIKE
Lanius excubitor
Wn 11–13 in **Wt** 1¾–3 oz
Breeds Alaska & N Canada; also Eurasia, May to Sep
Winters S Canada & N US, Oct to Apr
Journey length 300–3,400 miles

GRAY CATBIRD
Dumetella carolinensis
Wn 9–10 in **Wt** 1–2 oz
Breeds S Canada, US except SW, Apr to Sep
Winters SE US, W Indies & C America S to Panama, Oct to Mar
Journey length 185–4,000 miles

SAGE THRASHER
Oreoscoptes montanus
Wn 9–10 in **Wt** 1½–2½ oz
Breeds W US, Apr to Oct
Winters SW US, N Mexico, Oct to Mar
Journey length 85–2,500 miles

BROWN THRASHER
Toxostoma rufum
Wn 11–12 in **Wt** 1¾–2¾ oz
Breeds S Canada, C & E US, Apr to Oct
Winters E & S US, Oct to Mar
Journey length 0–1,500 miles

SPRAGUE'S PIPIT
Anthus spragueii
Wn 11–12 in **Wt** ¾ oz
Breeds Prairies of S Canada & N US, Apr to Sep
Winters CS US & Mexico, Oct to Apr
Journey length 1,250–3,000 miles

CEDAR WAXWING
Bombycilla cedrorum
Wn 10–12 in **Wt** 1¼–2 oz
Breeds S Canada & N US, Apr to Oct
Winters S US, Mexico & C America S to Panama, Nov to Mar
Journey length 0–3,400 miles

BLACK-CAPPED VIREO
Vireo atricapillus
Wn 6–9 in **Wt** ¼–½ oz
Breeds Oklahoma, Texas & NE Mexico, Mar to Aug
Winters C Mexico, Sep to Feb
Journey length 435–1,250 miles

YELLOW-THROATED VIREO
Vireo flavifrons
Wn 7–9 in **Wt** ½–¾ oz
Breeds SE Canada & E US, Apr to Oct
Winters SE US, Mexico, W Indies, C America & N S America, Nov to Mar
Journey length 300–4,000 miles

BELL'S VIREO *Vireo bellii*
Wn 7–9 in **Wt** ½ oz
Breeds C & W US, N Mexico, Mar to Sep
Winters Mexico & C America S to Nicaragua, Oct to Feb
Journey length 185–2,500 miles

GRAY VIREO *Vireo vicinior*
Wn 7–9 in **Wt** ½–¾ oz
Breeds SW US & N Mexico, Mar to Sep
Winters W Mexico, Oct to Feb
Journey length 155–1,400 miles

BLUE-HEADED VIREO *Vireo solitarius*
Wn 7–9 in **Wt** ½–¾ oz
Breeds S Canada, NE & W US, Mexico, Guatemala & El Salvador, Apr to Oct
Winters Extreme S US, Mexico, C America & W Indies, Oct to Mar
Journey length 600–4,700 miles

BLACK-WHISKERED VIREO *Vireo Altiloquus*
Wn 8–10 in **Wt** ¾–1 oz
Breeds Coastal Florida & W Indies, May to Sep
Winters S America S to Brazil & Peru, Oct to Apr
Journey length 1,250–2,800 miles

WARBLING VIREO *Vireo gilvus*
Wn 7–9 in **Wt** ½–¾ oz
Breeds SW Canada, US & N Mexico, Apr to Oct
Winters N Mexico, C America & S America S to Bolivia, Nov to Mar
Journey length 300–5,600 miles

PHILADELPHIA VIREO *Vireo Philadelphicus*
Wn 7–9 in **Wt** ½–¾ oz
Breeds S Canada, May to Sep
Winters Guatemala S to Panama & NW Colombia, Oct to Apr
Journey length 2,175–4,700 miles

PROTHONOTARY WARBLER
Protonotaria citrea
Wn 7–9 in **Wt** ½–¾ oz
Breeds E US, Apr to Sep
Winters W Indies & N S America, Oct to Mar
Journey length 900–3,000 miles

BLUE-WINGED WARBLER
Vermivora pinus
Wn 7–9 in **Wt** ¼–½ oz
Breeds E & NE US, Apr to Sep
Winters Mexico to Panama, Oct to Mar
Journey length 1,100–2,500 miles

GOLDEN-WINGED WARBLER
Vermivora chrysoptera
Wn 7–9 in **Wt** ¼–½ oz
Breeds SE Canada & NW US, Apr to Sep
Winters Guatemala to Colombia & Venezuela, Oct to Mar
Journey length 1,500–3,400 miles

TENNESSEE WARBLER
Vermivora peregrina
Wn 7–9 in **Wt** ¼–½ oz
Breeds S Canada & N US, May to Sep
Winters C Mexico to Venezuela & W Indies, Oct to Apr
Journey length 1,850–3,000 miles

ORANGE-CROWNED WARBLER
Vermivora celata
Wn 7–9 in **Wt** ½–¾ oz
Breeds Alaska, S Canada & W US, Apr to Sep
Winters Extreme S US, Mexico & Guatemala, Oct to Mar
Journey length 600–5,300 miles

NASHVILLE WARBLER
Vermivora ruficapilla
Wn 7–9 in **Wt** ¼–½ oz
Breeds SE Canada, NE & NW US, Apr to Sep
Winters Mexico & Guatemala, Oct to Mar
Journey length 300–3,000 miles

VIRGINIA'S WARBLER
Vermivora virgineae
Wn 6–8 in **Wt** ¼–½ oz
Breeds Mountains of W US, May to Sep
Winters Mexico, Oct to Apr
Journey length 1,250–2,175 miles

NORTHERN PARULA
Parula americana
Wn 6–8 in **Wt** ¼–½ oz
Breeds SE Canada & E US, Apr to Sep
Winters Florida, W Indies & Mexico S to Nicaragua, Oct to Mar
Journey length 300–3,000 miles

BLACK-AND-WHITE WARBLER
Mniotilta varia
Wn 6–8 in **Wt** ½ oz
Breeds S Canada & E US, Apr to Sep
Winters SE US, Mexico, C America, W Indies & N S America, Oct to Mar
Journey length 300–4,700 miles

BLACK-THROATED BLUE WARBLER
Dendroica caerulescens
Wn 6–8 in **Wt** ½ oz
Breeds SE Canada & NE US, May to Sep
Winters W Indies, C America & N S America, Oct to Apr
Journey length 1,500–4,000 miles

CERULEAN WARBLER
Dendroica cerulea
Wn 6–7 in **Wt** ¼–½ oz
Breeds SE Canada & E US, Apr to Sep
Winters S America from Colombia & Venezuela to Bolivia, Oct to Mar
Journey length 2,175–4,350 miles

BLACKBURNIAN WARBLER
Dendroica fusca
Wn 6–7 in **Wt** ½ oz
Breeds S & E Canada & NE US, May to Sep
Winters Costa Rica, Panama & S America S to Peru, Oct to Apr
Journey length 1,500–4,700 miles

CHESTNUT-SIDED WARBLER
Dendroica pensylvanica
Wn 6–8 in **Wt** ¼–½ oz
Breeds SE Canada & NE US, May to Sep
Winters Guatemala to Colombia, Oct to Apr
Journey length 1,850–4,000 miles

CAPE MAY WARBLER
Dendroica tigrina
Wn 6–8 in **Wt** ¼–½ oz
Breeds S Canada & NE US, May to Sep
Winters W Indies, Oct to Apr
Journey length 1,500–2,500 miles

MAGNOLIA WARBLER
Dendroica magnolia
Wn 6–8 in **Wt** ¼–½ oz
Breeds S Canada & NE US, May to Sep
Winters W Indies, C America & N S America, Oct to Apr
Journey length 1,500–4,000 miles

YELLOW-RUMPED WARBLER
Dendroica coronata
Wn 7–9 in **Wt** ½ oz
Breeds Alaska, Canada & W US, Apr to Sep
Winters S US, Mexico, C America, W Indies & N S America, Oct to Mar
Journey length 300–6,000 miles

BLACK-THROATED GRAY WARBLER
Dendroica nigrescens
Wn 6–8 in **Wt** ½ oz
Breeds W Canada, W US & NW Mexico, Apr to Sep
Winters Mexico & Guatemala, Oct to Mar
Journey length 155–3,750 miles

TOWNSEND'S WARBLER
Dendroica townsendi
Wn 6–8 in **Wt** ¼–½ oz
Breeds S Alaska, W Canada & NW US, Apr to Sep
Winters W California, Mexico & C America S to Nicaragua, Oct to Mar
Journey length 125–4,000 miles

HERMIT WARBLER
Dendroica occidentalis
Wn 7–9 in **Wt** ½ oz
Breeds N California & Oregon, May to Sep
Winters SW California, Mexico & C America S to Nicaragua, Oct to Apr
Journey length 185–3,400 miles

BLACK-THROATED GREEN WARBLER
Dendroica virens
Wn 6–8 in **Wt** ¼–½ oz
Breeds E & S Canada & NE US, Apr to Sep
Winters Mexico & W Indies, Oct to Mar
Journey length 1,100–2,800 miles

GOLDEN-CHEEKED WARBLER
Dendroica chrysoparia
Wn 7–9 in **Wt** ½ oz
Breeds Texas, Apr to Sep
Winters Mexico S to Nicaragua, Oct to Mar
Journey length 900–1,500 miles

GRACE'S WARBLER
Dendroica graciae
Wn 6–8 in **Wt** ¼–½ oz
Breeds SW US & N Mexico, Apr to Sep
Winters Mexico S to Nicaragua, Oct to Mar
Journey length 155–1,250 miles

PRAIRIE WARBLER
Dendroica discolor
Wn 6–7 in **Wt** ¼–½ oz
Breeds SE Canada & E US, Apr to Oct
Winters Florida, W Indies & C America S to Nicaragua, Oct to Mar
Journey length 470–2,500 miles

BAY-BREASTED WARBLER
Dendroica castanea
Wn 7–9 in **Wt** ½ oz
Breeds S Canada & NE US, May to Sep
Winters Panama to Colombia & Venezuela, Oct to Apr
Journey length 2,175–3,400 miles

PINE WARBLER
Dendroica pinus
Wn 7–9 in **Wt** ½ oz
Breeds SE Canada & E US Apr to Oct
Winters SE US, Mexico & W Indies, Oct to Mar
Journey length 0–2,500 miles

PALM WARBLER
Dendroica palmarum
Wn 7–9 in **Wt** ½ oz
Breeds S Canada & NE US, May to Sep
Winters SE US, Mexico S to Honduras & W Indies, Oct to Apr
Journey length 600–3,000 miles

MOURNING WARBLER
Oporornis philadelphia
Wn 6–8 in **Wt** ½ oz
Breeds S Canada & NE US, May to Sep
Winters S C America & N S America, Oct to Apr
Journey length 1,500–4,350 miles

MACGILLIVRAY'S WARBLER
Oporornis tolmiei
Wn 6–8 in **Wt** ¼–½ oz
Breeds W Canada & W US, Apr to Sep
Winters Mexico to Panama, Oct to Mar
Journey length 1,250–5,000 miles

CONNECTICUT WARBLER
Oporornis agilis
Wn 6–9 in **Wt** ½ oz
Breeds S Canada & N US, May to Sep
Winters C & S America S to Brazil, Oct to Apr
Journey length 2,500–5,600 miles

KENTUCKY WARBLER
Oporornis formosus
Wn 6–8 in **Wt** ¼–½ oz
Breeds E US, Apr to Oct
Winters C & N S America, Nov to Mar
Journey length 1,250–5,000 miles

WILSON'S WARBLER
Wilsonia pusilla
Wn 5–7 in **Wt** ¼–½ oz
Breeds Alaska, Canada & W US, May to Sep
Winters Mexico to Panama, Oct to Apr
Journey length 2,500–5,600 miles/

HOODED WARBLER
Wilsonia citrina
Wn 6–8 in **Wt** ½ oz
Breeds SE Canada & E US, Apr to Sep
Winters Mexico to Panama, Oct to Mar
Journey length 2,500–4,350 miles

OVENBIRD
Ceryle alcyon
Wn 9–10 in **Wt** ½–¾ oz
Breeds S Canada, C & E US, Apr to Oct
Winters Florida, Mexico to N S America & W Indies, Nov to Mar
Journey length 600–4,350 miles

NORTHERN WATERTHRUSH
Seiurus noveboracensis
Wn 9–10 in **Wt** ½–¾ oz
Breeds Alaska, Canada & N US, May to Oct
Winters Florida, Mexico & W Indies S to Peru, Nov to Apr
Journey length 1,850–6,000 miles

AMERICAN REDSTART
Setophaga ruticilla
Wn 6–8 in **Wt** ½ oz
Breeds S Canada, C & N US, Apr to Sep
Winters S Mexico S to Ecuador & W Indies, Oct to Mar
Journey length 1,500–4,700 miles

ROSE-BREASTED GROSBEAK
Pheucticus ludovicianus
Wn 9–12 in **Wt** 1¼–2 oz
Breeds S Canada & NE US, May to Oct
Winters C America & N S America, Nov to Apr
Journey length 2,500–5,000 miles

INDIGO BUNTING
Passerina cyanea
Wn 6–9 in **Wt** ½–¾ oz
Breeds C & E US, Apr to Sep
Winters Mexico to Panama & W Indies, Oct to Mar
Journey length 470–3,750 miles

LAZULI BUNTING
Passerina amoena
Wn 6–9 in **Wt** ½–¾ oz
Breeds SW Canada & W US, Apr to Oct
Winters Mexico, Nov to Mar
Journey length 600–3,400 miles

PAINTED BUNTING *Passerina ciris*
Wn 6–9 in **Wt** ½–¾ oz
Breeds S US & N Mexico, Mar to Sep
Winters Mexico to Panama & Cuba, Oct to Feb
Journey length 300–3,000 miles

GREEN-TAILED TOWHEE
Pipilo chlorurus
Wn 10–12 in **Wt** 1–1¼ oz
Breeds W US, Apr to Sep
Winters SW US & Mexico, Oct to Mar
Journey length 300–3,000 miles

BAIRD'S SPARROW
Ammodramus bairdii
Wn 6–9 in **Wt** ½–¾ oz
Breeds Prairies of S Canada & N US, May to Sep
Winters N Mexico, Oct to Apr
Journey length 1,500–2,500 miles

AMERICAN TREE SPARROW
Spizella arborea
Wn 8–9 in **Wt** ¾–1 oz
Breeds Alaska & N Canada, May to Sep
Winters US, Oct to Apr
Journey length 1,500–3,750 miles

MCCOWN'S LONGSPUR
Calcarius mccownii
Wn 6–9 in **Wt** ½–¾ oz
Breeds Prairies of S Canada & N US, Apr to Oct
Winters Texas, Arizona & N Mexico, Nov to Mar
Journey length 600–2,800 miles

DICKCISSEL *Spiza americana*
Wn 8–9 in **Wt** ½–¾ oz
Breeds S Canada, C & E US, Apr to Sep
Winters S Mexico S to N S America, Oct to Mar
Journey length 900–4,700 miles

RUSTY BLACKBIRD
Euphagus carolinus
Wn 12–14 in **Wt** 2–3 oz
Breeds Alaska, Canada & NE US, May to Oct
Winters E & SE US, Nov to Apr
Journey length 300–4,000 miles

SUMMER TANAGER *Piranga rubra*
Wn 11–13 in **Wt** 2–2½ oz
Breeds S US & N Mexico, Mar to Sep
Winters S Mexico, W Indies & S to Brazil, Oct to Feb
Journey length 600–4,350 miles

EVENING GROSBEAK
Coccothraustes vespertinus
Wn 11–14 in **Wt** 1¾–3 oz
Breeds S Canada, N & W US, Mar to Oct
Winters Breeding range plus S & E US, Nov to Feb
Journey length 0–1,250 miles

EURASIAN MIGRANTS

RED-NECKED GREBE
Podiceps grisegena
Wn 30–33 in **Wt** 14 oz–2 lb 2 oz
Breeds N Europe, Asia & N America, Apr to Sep
Winters Coasts of N America, E Asia & NW Europe, Oct to Mar
Journey length 155–2,500 miles

HORNED GREBE *Podiceps auritus*
Wn 23–25 in **Wt** 11 oz–1 lb 1 oz
Breeds N Europe, Asia & N America, Apr to Sep
Winters Coasts of N America, E Asia & NW Europe, Oct to Mar
Journey length 155–3,000 miles

WHITE PELICAN
Pelecanus onocrotalus
Wn 106–141 in **Wt** 20 lb–24 lb 4 oz
Breeds Scattered waters from Greece to Mongolia; resident in Africa & India, Apr to Sep
Winters Red Sea S to E Africa, Oct to Mar
Journey length 900–4,000 miles

LITTLE BITTERN
Ixobrychus minutus
Wn 20–22 in **Wt** 4–6 oz
Breeds C & S Europe, SW Asia; resident in Africa, India & Australia, Apr to Sep
Winters E Africa S to S Africa, Oct to Mar
Journey length 1,500–6,800 miles

SQUACCO HERON *Ardeola ralloides*
Wn 31–36 in **Wt** 7–13 oz
Breeds S Europe & SW Asia; resident in Africa, Apr to Sep
Winters W, E & S Africa, Nov to Mar
Journey length 2,500–5,300 miles

LITTLE EGRET *Egretta garzetta*
Wn 34–37 in **Wt** 14 oz–1 lb 6 oz
Breeds S Europe, N Africa, SW Asia; resident in S Africa, India, SE Asia & Australasia, Mar to Sep
Winters N, W & E Africa, Oct to Mar
Journey length 300–4,000 miles

PURPLE HERON *Ardea purpurea*
Wn 47–59 in
Wt 1 lb 3 oz–2 lb 11 oz
Breeds S Europe, SW Asia & China; resident in E Africa, India, SE Asia, Apr to Sep
Winters W & E Africa, SE Asia, Oct to Mar
Journey length 1,850–4,350 miles

GLOSSY IBIS *Plegadis falcinellus*
Wn 31–37 in
Wt 1 lb 3 oz–1 lb 11 oz
Breeds SE & SW Europe; resident in E & S Africa, India & Australia, Apr to Sep
Winters W & E Africa, Oct to Mar
Journey length 2,175–2,800 miles

HERMIT IBIS *Geronticus eremita*
Wn 49–53 in
Wt 2 lb 9 oz–3 lb 3 oz
Breeds Morocco & Turkey, Feb to Aug
Winters Morocco, Mauritania & NE Africa, Sep to Jan
Journey length 300–1,850 miles

EURASIAN SPOONBILL
Platalea leucorodia
Wn 45–51 in
Wt 2 lb 14 oz–3 lb 12 oz
Breeds W & S Europe, C & S Asia; resident in India, Apr to Sep
Winters N & NE Africa, SE China, Oct to Mar
Journey length 155–2,175 miles

BEAN GOOSE *Anser fabalis*
Wn 55–68 in
Wt 4 lb 7 oz–8 lb 15 oz
Breeds N Europe & Asia, May to Sep
Winters W Europe, Japan & China, Oct to Apr
Journey length 600–2,800 miles

EURASIAN WIGEON *Anas penelope*
Wn 29–33 in **Wt** 14 oz–2 lb 6 oz
Breeds N Europe & Asia, May to Sep
Winters W Europe, NE Africa, India, Japan & SE Asia, Oct to Apr
Journey length 300–4,700 miles

NORTHERN PINTAIL *Anas acuta*
Wn 31–37 in **Wt** 1 lb 3 oz–2 lb 14 oz
Breeds N Europe, Asia & N N America, May to Sep
Winters Europe, S Asia, W & E Africa, N & C America, Oct to Apr
Journey length 300–5,600 miles

GARGANEY *Anas querquedula*
Wn 23–24 in **Wt** 9 oz–1 lb 5 oz
Breeds W & C Europe & Asia, May to Sep
Winters W & E Africa, India & SE Asia, Oct to Mar
Journey length 2,175–4,350 miles

NORTHERN SHOVELER
Anas clypeata
Wn 27–33 in **Wt** 12 oz–2 lb 3 oz
Breeds W & N Europe, N Asia, Apr to Sep
Winters Europe, E Africa, India, SE Asia & Japan, Oct to Mar
Journey length 60–4,700 miles

RED-CRESTED POCHARD
Netta rufina
Wn 33–34 in **Wt** 1 lb 13 oz–3 lb 2 oz
Breeds S Europe, SW & C Asia, Apr to Sep
Winters Mediterranean countries, Indian subcontinent, Oct to Mar
Journey length 60–2,175 miles

FERRUGINOUS DUCK
Aythya nyroca
Wn 24–26 in **Wt** 1 lb–1 lb 10 oz
Breeds C & S Europe, SW Asia, Mar to Sep
Winters Mediterranean countries, NE Africa, N Pakistan & India, Oct to Mar
Journey length 60–2,800 miles

TUFTED DUCK *Aythya fuligula*
Wn 26–28 in **Wt** 14 oz–2 lb 4 oz
Breeds N Europe & Asia, Apr to Sep
Winters W & S Europe, S Asia & Japan, Oct to Mar
Journey length 60–4,000 miles

HARLEQUIN DUCK
Histrionicus histrionicus
Wn 24–27 in
Wt 1 lb 2 oz–1 lb 10 oz
Breeds NW & NE N America, Iceland, Greenland & NE Asia, May to Sep
Winters Coastal areas of breeding range, Oct to Apr
Journey length 30–1,250 miles

COMMON GOLDENEYE
Bucephala clangula
Wn 25–31 in
Wt 1 lb 2 oz–2 lb 12 oz
Breeds N Europe, Asia & N America, May to Sep
Winters W & S Europe, SE Asia & US, Oct to Apr
Journey length 300–1,850 miles

COMMON MERGANSER
Mergus merganser
Wn 32–38 in **Wt** 2 lb–4 lb 12 oz
Breeds N Europe, Asia & N America; resident in SC Asia, Apr to Sep
Winters W Europe, E Asia & US, Oct to Apr
Journey length 60–1,850 miles

BLACK KITE *Milvus migrans*
Wn 62–70 in
Wt 1 lb 6 oz–2 lb 1 oz
Breeds Eurasia except extreme N, resident in Indian subcontinent, SE Asia, Africa & Australasia, Apr to Sep
Winters Africa, India, SE Asia, Oct to Mar
Journey length 600–2,800 miles

EGYPTIAN VULTURE
Neophron percnopterus
Wn 61–70 in
Wt 3 lb 8 oz–4 lb 13 oz
Breeds S Europe, N Africa & SW Asia; resident in W & E Africa & Indian subcontinent, Apr to Sep
Winters W & E Africa, Arabian peninsula & Indian subcontinent, Oct to Mar
Journey length 600–1,850 miles

SHORT-TOED EAGLE
Circaetus gallicus
Wn 72–76 in
Wt 2 lb 15 oz–5 lb 2 oz
Breeds C & S Europe, SW Asia; resident in Indian subcontinent, Oct to Mar
Winters N tropics of Africa & Indian subcontinent, Mar to Oct
Journey length 600–3,750 miles

PALLID HARRIER *Circus macrourus*
Wn 37–47 in **Wt** 8 oz–1 lb 5 oz
Breeds CE Europe & CW Asia, Apr to Oct
Winters Tropical Africa S to Cape, Indian subcontinent, Oct to Apr
Journey length 2,175–7,000 miles

MONTAGU'S HARRIER
Circus pygargus
Wn 41–47 in **Wt** 8 oz–1 lb
Breeds S & C Europe & CW Asia, Apr to Sep
Winters Tropical Africa S to Cape, Indian subcontinent, Oct to Mar
Journey length 2,175–7,150 miles

LEVANT SPARROWHAWK
Accipiter brevipes
Wn 25–29 in **Wt** 5–10 oz
Breeds SE & CE Europe, Apr to Aug
Winters NE Africa, Sep to Apr
Journey length 900–3,000 miles

LESSER SPOTTED EAGLE
Aquila pomarina
Wn 52–62 in
Wt 2 lb 6 oz–4 lb 12 oz
Breeds E & SE Europe; resident in Indian subcontinent, Apr to Oct
Winters E & SE Africa, Oct to Mar
Journey length 1,850–6,400 miles

SPOTTED EAGLE *Aquila clanga*
Wn 61–71 in **Wt** 3 lb–7 lb 1 oz
Breeds E Europe & forests of Russia, Mongolia & N China, Apr to Sep
Winters SE Europe, Indian subcontinent & SE Asia, Oct to Mar
Journey length 600–4,350 miles

BOOTED EAGLE
Hieraaetus pennatus
Wn 39–47 in
Wt 1 lb 2 oz–2 lb 12 oz
Breeds S Europe, SW & C Asia, Apr to Oct
Winters Tropical Africa S to Cape & Indian subcontinent, Oct to Mar
Journey length 2,175–6,800 miles

EURASIAN HOBBY *Falco subbuteo*
Wn 32–36 in **Wt** 5–12 oz
Breeds Eurasia except extreme N & S, Apr to Sep
Winters S Africa, Indian subcontinent & SE Asia, Oct to Mar
Journey length 600–7,150 miles

SOOTY FALCON *Falco concolor*
Wn 33–43 in **Wt** 11–16 oz
Breeds NE Africa & Arabia, May to Oct
Winters Madagascar & Mozambique, Nov to May
Journey length 1,500–4,000 miles

SPOTTED CRAKE *Porzana porzana*
Wn 14–16 in **Wt** 2–5 oz
Breeds W Europe except extreme N, S & W Asia, Apr to Aug
Winters E & SE Africa & Indian subcontinent, Sep to Mar
Journey length 1,850–7,000 miles

LITTLE CRAKE *Porzana parva*
Wn 13–15 in **Wt** 1–3 oz
Breeds S and C Europe, CW Asia, Apr to Sep
Winters E Africa & Pakistan, Oct to Mar
Journey length 1,250–4,475 miles

BAILLION'S CRAKE
Porzana pusilla
Wn 12–14 in **Wt** 1–2 oz
Breeds S & C Europe, middle latitudes of Asia; resident in S Africa & Australasia, Apr to Oct
Winters E Africa, Indian subcontinent & SE Asia, Oct to Apr
Journey length 600–4,700 miles

MANCHURIAN CRANE
Grus japonensis
Wn 90–98 in
Wt 13 lb 4 oz–19 lb 13 oz
Breeds E Russia & NE China; resident in Japan, Apr to Oct
Winters Korea & N China, Nov to Mar
Journey length 600–1,250 miles

COLLARED PRATINCOLE
Glareola pratincola
Wn 23–25 in **Wt** 2–3 oz
Breeds S Europe & SW Asia; resident in Africa, Apr to Sep
Winters Sub-Saharan Africa, Oct to Mar
Journey length 900–3,750 miles

BLACK-WINGED PRATINCOLE
Glareola nordmanni
Wn 23–26 in **Wt** 3–4 oz
Breeds CE Europe & W Asia, Apr to Sep
Winters W & S Africa, Oct to Apr
Journey length 1,750–6,000 miles

LITTLE PLOVER *Charadrius dubius*
Wn 16–18 in **Wt** 1–1¾ oz
Breeds Eurasia except extreme N; resident in Indian subcontinent & SE Asia, Mar to Sep
Winters Tropical Africa, Indian subcontinent, SE Asia & Australasia, Sep to Mar
Journey length 600–5,300 miles

MONGOLIAN PLOVER
Charadrius mongolus
Wn 17–22 in **Wt** 1½–2½ oz
Breeds E Siberia, W China, Himalayas, May to Sep
Winters Coasts of E Africa, Arabia, Indian subcontinent & Australasia, Oct to Apr
Journey length 1,250–8,000 miles

GREATER SAND PLOVER
Charadrius leschenaultii
Wn 20–23 in **Wt** 3–4 oz
Breeds SE Europe & C Asia, Apr to Aug
Winters Coasts of E Africa, Arabia, Indian subcontinent & Australasia, Aug to Apr
Journey length 900–6,000 miles

CASPIAN PLOVER
Charadrius asiaticus
Wn 21–24 in **Wt** 2–3 oz
Breeds SW Asia, Apr to Sep
Winters E & S Africa, Oct to Mar
Journey length 1,500–6,700 miles

SOCIABLE PLOVER
Chettusia gregaria
Wn 27–29 in **Wt** 6–9 oz
Breeds CE Europe & W Asia, Apr to Oct
Winters NE Africa, Iraq & Pakistan, Oct to Apr
Journey length 1,250–3,400 miles

JACK SNIPE
Lymnocryptes minimus
Wn 14–16 in **Wt** 1¼–3¼ oz
Breeds N Eurasia, Mar to Oct
Winters W & S Europe, N & tropical Africa, Middle East & Indian subcontinent to Vietnam, Oct to Mar
Journey length 300–5,000 miles

GREAT SNIPE *Gallinago media*
Wn 18–19 in **Wt** 5–9 oz
Breeds N Europe & NW Russia, Apr to Oct
Winters Tropical & S Africa, Oct to Mar
Journey length 2,175–8,200 miles

PINTAIL SNIPE
Gallinago stenura
Wn 17–18 in **Wt** 3–5 oz
Breeds N & C Asia, May to Sep
Winters Indian subcontinent, SE Asia & Indonesia, Sep to Apr
Journey length 1,500–6,000 miles

BLACK-TAILED GODWIT
Limosa limosa
Wn 27–32 in **Wt** 6 oz–1 lb 2 oz
Breeds Temperate Europe, W & E Asia, Mar to Aug
Winters Mediterranean, tropical Africa, Indian subcontinent, SE Asia & Australasia, Sep to Mar
Journey length 900–7,800 miles

WHIMBREL *Numenius phaeopus*
Wn 29–35 in **Wt** 10 oz–1 lb 5 oz
Breeds N Eurasia, Alaska & CN Canada, Apr to Oct
Winters Coasts of Africa, Arabia, Australasia, N, C & S America, Oct to Apr
Journey length 1,500–9,300 miles

SPOTTED REDSHANK
Tringa erythropus
Wn 24–26 in **Wt** 4–7 oz
Breeds N Eurasia, May to Aug
Winters Mediterranean, tropical Africa, Indian subcontinent & SE Asia, Sep to Apr
Journey length 1,250–5,300 miles

MARSH SANDPIPER
Tringa stagnatilis
Wn 21–23 in **Wt** 2–4 oz
Breeds CE Europe, W & C Asia, Apr to Sep
Winters Tropical & S Africa, Arabia, Indian subcontinent & Australasia, Oct to Mar
Journey length 1,850–7,500 miles

COMMON GREENSHANK
Tringa nebularia
Wn 26–27 in **Wt** 5–10 oz
Breeds N Eurasia, Apr to Oct
Winters Tropical & S Africa, Arabia, Indian subcontinent, SE Asia & Australasia, Oct to Apr
Journey length 1,500–7,500 miles

GREEN SANDPIPER
Tringa ochropus
Wn 22–24 in **Wt** 2–4 oz
Breeds N Eurasia, Apr to Aug
Winters W & S Europe, tropical Africa, Middle East, Indian subcontinent & SE Asia, Sep to Mar
Journey length 300–5,600 miles

WOOD SANDPIPER *Tringa glareola*
Wn 22–22 in **Wt** 1¾–3½ oz
Breeds N Eurasia, Apr to Aug
Winters Tropical & S Africa, Indian subcontinent, SE Asia & Australasia, Sep to Apr
Journey length 1,500–8,000 miles

TEREK SANDPIPER *Xenus cinereus*
Wn 22–23 in **Wt** 2–4 oz
Breeds NE Europe & N Asia, May to Aug
Winters Coasts of Africa, Arabia, Indian subcontinent, SE Asia & Australasia, Sep to Apr
Journey length 1,850–8,000 miles

COMMON SANDPIPER
Actitis hypoleucos
Wn 14–16 in **Wt** 1¼–2½ oz
Breeds Eurasia except S, Apr to Aug
Winters Tropical & S Africa, Indian subcontinent, SE Asia & Australasia, Sep to Apr
Journey length 300–8,000 miles

GREAT BLACK-HEADED GULL
Larus ichthyaetus
Wn 58–66 in
Wt 2 lb 2 oz–4 lb 7 oz
Breeds CE Europe & C Asia, Apr to Oct
Winters Caspian Sea, Red Sea, Gulf & coasts of Indian subcontinent, Oct to Apr
Journey length 155–3,750 miles

MEDITERRANEAN GULL
Larus melanocephalus
Wn 36–39 in **Wt** 9–14 oz
Breeds S & W Europe, Apr to Sep
Winters Coasts of W Europe, Black & Mediterranean seas, Sep to Apr
Journey length 155–1,850 miles

BLACK-HEADED GULL
Larus ridibundus
Wn 39–43 in **Wt** 6–12 oz
Breeds Eurasia; resident in W Europe, Apr to Sep
Winters Coasts & offshore E N America, Europe, N Africa, Middle East & E Asia, Oct to Mar
Journey length-300–4,700 miles

LITTLE GULL
Larus minutus
Wn 29–31 in **Wt** 3–5 oz
Breeds NC Europe & NC Asia, Apr to Oct
Winters Coasts of Europe & Caspian Sea, Oct to Apr
Journey length 300–2,175 miles

GULL-BILLED TERN
Gelochelidon nilotica
Wn 39–45 in **Wt** 7–10 oz
Breeds S Europe, S & C Asia, E & W coasts of N America, Apr to Sep
Winters W & S Europe, tropical Africa, Arabia, S Asia, Australasia,

C & S America, Oct to Mar
Journey length 300–6,000 miles

SANDWICH TERN
Sterna sandvicensis
Wn 37–41 in **Wt** 8–10 oz
Breeds NW Europe, Mediterranean, Black & Caspian seas; also SE US, W Indies & S America, Apr to Oct
Winters Coasts of W & S Europe, Black & Caspian seas, Arabia, W & S Africa; Gulf of Mexico & S America, Oct to Apr
Journey length 300–6,800 miles

COMMON TERN
Sterna hirundo
Wn 30–38 in **Wt** 4–6 oz
Breeds Eurasia except extreme N & S, Canada & E US, Apr to Sep
Winters Coasts & offshore Eurasia, Africa, Australasia, N & S America, Sep to Mar
Journey length 600–8,000 miles

WHITE-CHEEKED TERN
Sterna repressa
Wn 29–32 in **Wt** 4–5 oz
Breeds Red Sea, Gulf, E Africa & W India, Apr to Oct
Winters Coastal areas to S of breeding range, Oct to Apr
Journey length 300–1,250 miles

WHISKERED TERN
Chlidonias hybridus
Wn 29–30 in **Wt** 3–4 oz
Breeds S Europe, SW & SE Asia; resident in SE Africa, India & Australia, May to Oct
Winters Tropical Africa, Indian subcontinent & Indonesia, Oct to Apr
Journey length 1,850–5,000 miles

WHITE-WINGED BLACK TERN
Chlidonias leucopterus
Wn 24–26 in **Wt** 2–3 oz
Breeds CE Europe, W & E Asia, May to Sep
Winters Tropical & S Africa, SE Asia & Australasia, Oct to Apr
Journey length 2,175–5,600 miles

STOCK PIGEON
Columba oenas
Wn 24–27 in **Wt** 9–13 oz
Breeds Europe except extreme N, SW Asia; resident in W Europe & parts of SW Asia, Mar to Oct
Winters W & S Europe, SW Asia, Mar to Oct
Journey length 60–2,500 miles

YELLOW-EYED STOCK PIGEON
Columba eversmanni
Wn 23–24 in **Wt** 6–8 oz
Breeds Iran & Afghanistan N to Kazakhstan, Apr to Oct
Winters NW India & Pakistan, Oct to Apr
Journey length 300–1,250 miles

COMMON WOOD PIGEON
Columba palumbus
Wn 29–31 in **Wt** 11 oz–1 lb 6 oz
Breeds Europe, N Africa & W Asia; resident in W Europe & N Africa, Apr to Oct
Winters W & S Europe & SW Asia, Oct to Apr
Journey length 60–2,175 miles

RUFOUS TURTLE DOVE
Streptopelia orientalis
Wn 20–23 in **Wt** 6–10 oz
Breeds E & S Asia; resident in S of breeding range, May to Oct
Winters India, Burma, Japan & China, Oct to Apr
Journey length 300–2,500 miles

GREAT SPOTTED CUCKOO
Clamator glandarius
Wn 22–24 in **Wt** 5–7 oz
Breeds S Europe, S Africa; resident in tropical Africa, May to Sep
Winters Tropical Africa, Oct to Apr
Journey length 1,850–2,800 miles

ORIENTAL CUCKOO
Cuculus saturatus
Wn 20–22 in **Wt** 3–5 oz
Breeds Asia except C & SW; resident in Indonesia, May to Aug
Winters SE Asia, Indonesia & Australia, Sep to Apr
Journey length 300–7,150 miles

EUROPEAN NIGHTJAR
Caprimulgus europaeus
Wn 22–25 in **Wt** 2–4 oz
Breeds Europe & W Asia, Apr to Sep
Winters E & S Africa, Oct to Apr
Journey length 1,850–7,500 miles

RED-NECKED NIGHTJAR
Caprimulgus ruficollis
Wn 25–26 in **Wt** 2–4 oz
Breeds Iberia & NW Africa, Apr to Oct
Winters W Africa, Oct to Mar
Journey length 1,100–2,000 miles

EGYPTIAN NIGHTJAR
Caprimulgus aegyptius
Wn 22–26 in **Wt** 2–3 oz
Breeds NW & NE Africa, Kazakhstan, Turkmenistan & Iraq, Mar to Oct
Winters Tropical W & E Africa, Oct to Mar
Journey length 900–3,400 miles

PALLID SWIFT
Apus pallidus
Wn 16–18 in **Wt** 1–1¾ oz
Breeds S Europe, N Africa & Gulf, Mar to Oct
Winters Tropical Africa, Oct to Feb
Journey length 1,100–3,400 miles

ALPINE SWIFT
Apus melba
Wn 21–23 in **Wt** 3–4 oz
Breeds S Europe, N Africa, Middle East & SW Asia; resident in India & E & S Africa, Apr to Oct
Winters E & S Africa, Oct to Apr
Journey length 600–5,600 miles

BLUE-CHEEKED BEE-EATER
Merops superciliosus
Wn 18–19 in **Wt** 1½–2 oz
Breeds N Africa, Middle East, SW Asia & Pakistan; scattered resident in tropical & S Africa, May to Sep
Winters W Africa, SE Africa & Pakistan, Oct to Apr
Journey length 600–2,500 miles

SHORT-TOED LARK
Calandrella brachydactyla
Wn 9–11 in **Wt** ½–1 oz
Breeds S Europe, N Africa, Middle East & SW Asia, Mar to Sep
Winters Tropical Africa, SW Asia, Oct to Mar
Journey length 300–2,800 miles

RED-RUMPED SWALLOW
Hirundo daurica
Wn 12–13 in **Wt** ¾–1 oz
Breeds S Europe, N Africa, S Asia; resident in Indian subcontinent & Africa, Apr to Sep
Winters Tropical Africa, Indian subcontinent & Burma, Sep to Apr
Journey length 600–3,000 miles

HOUSE MARTIN
Delichon urbica
Wn 10–11 in **Wt** ½–¾ oz
Breeds Eurasia & N Africa, Apr to Oct
Winters Sub-Saharan Africa, India, SE Asia & Philippines, Oct to Apr
Journey length 900–8,000 miles

BLYTH'S PIPIT
Anthus godlewskii
Wn 11–12 in **Wt** ¾–1 oz
Breeds C Asia & Mongolia S to Tibet, May to Aug
Winters India & Sri Lanka, Sep to Apr
Journey length 900–3,000 miles

TAWNY PIPIT
Anthus campestris
Wn 9–11 in **Wt** ½–1 oz
Breeds S & C Europe, N Africa, W & C Asia, Apr to Sep
Winters Tropical Africa, Arabia & Indian subcontinent, Sep to Mar
Journey length 900–4,475 miles

OLIVE-TREE PIPIT
Anthus hodgsoni
Wn 9–10 in **Wt** ½–1 oz
Breeds N & E Asia, May to Sep
Winters Indian subcontinent, SE Asia & Philippines, Oct to Apr
Journey length 300–5,600 miles

TREE PIPIT
Anthus trivialis
Wn 9–10 in **Wt** ½–1 oz
Breeds Europe, W & C Asia, May to Sep
Winters Tropical & S Africa & Indian subcontinent, Oct to Apr
Journey length 600–6,500 miles

PECHORA PIPIT
Anthus gustavi
Wn 9–10 in **Wt** ½–1 oz
Breeds N Asia, Jun to Sep
Winters Philippines & Indonesia, Oct to Apr
Journey length 5,000–5,600 miles

MEADOW PIPIT
Anthus pratensis
Wn 8–9 in **Wt** ½–¾ oz
Breeds W, N & C Europe, Iceland, Apr to Oct
Winters S & W Europe, N Africa & SW Asia, Oct to Apr
Journey length 60–3,000 miles

CITRINE WAGTAIL
Motacilla citreola
Wn 9–10 in **Wt** ½–¾ oz
Breeds C Europe & Asia except extreme E, May to Sep
Winters Indian subcontinent & SE Asia, Oct to Apr
Journey length 300–4,700 miles

GRAY WAGTAIL
Motacilla cinerea
Wn 9–10 in **Wt** ½–1 oz
Breeds Europe except N & E, Asia; resident in W Europe, Apr to Oct
Winters W & S Europe, E Africa, Indian subcontinent, SE Asia & Australasia, Oct to Apr
Journey length 0–5,300 miles

WHITE WAGTAIL *Motacilla alba*
Wn 9–11 in **Wt** ½–1 oz
Breeds Eurasia; resident in W & S Europe & S Asia, Apr to Oct
Winters N & tropical Africa, Arabia, Indian subcontinent & SW Asia, Oct to Apr
Journey length 0–5,600 miles

SIBERIAN ACCENTOR
Prunella montanella
Wn 8–9 in **Wt** ½–¾ oz
Breeds N Asia, May to Sep
Winters N China, Oct to Apr
Journey length 1,850–2,500 miles

RUFOUS-TAILED SCRUB ROBIN
Cercotrichas galactotes
Wn 8–10 in **Wt** ¾ oz
Breeds SW & SE Europe, SW & N Africa; resident in tropical Africa, Apr to Sep
Winters Tropical W & E Africa, Sep to Apr
Journey length 600–3,400 miles

THRUSH NIGHTINGALE
Luscinia luscinia
Wn 9–10 in **Wt** ¾–1 oz
Breeds C Europe, W Asia, Apr to Sep
Winters E & SE Africa, Oct to Mar
Journey length 2,500–6,500 miles

SIBERIAN RUBYTHROAT
Luscinia calliope
Wn 9–10 in **Wt** ½–1 oz
Breeds N & E Asia, May to Aug
Winters SE Asia & Philippines, Sep to Apr
Journey length 900–5,000 miles

SIBERIAN BLUE ROBIN
Luscinia cyane
Wn 7–8 in **Wt** ½ oz
Breeds S Siberia S & E to Korea & Japan, May to Aug
Winters SE Asia & Philippines, Sep to Apr
Journey length 1,250–3,400 miles

RED-FLANKED BLUETAIL
Tarsiger cyanurus
Wn 8–9 in **Wt** ½ oz
Breeds E Asia & Himalayas, May to Sep
Winters S China & SE Asia, Oct to Apr
Journey length 600–3,400 miles

WHITE-THROATED ROBIN
Irania gutturalis
Wn 10–11 in **Wt** ½–1 oz
Breeds Turkey & SW Asia, May to Sep
Winters Kenya & Tanzania, Oct to Apr
Journey length 1,250–3,400 miles

WHINCHAT
Saxicola rubetra
Wn 8–9 in **Wt** ½–¾ oz
Breeds Europe & extreme W Asia, Apr to Sep
Winters Tropical & E Africa, Oct to Mar
Journey length 1,850–6,000 miles

COMMON STONECHAT
Saxicola torquata
Wn 7–8 in **Wt** ½–¾ oz
Breeds Eurasia; resident in W Europe, S & E Africa, Apr to Oct
Winters N & E Africa, Middle East, Indian subcontinent & SE Asia, Oct to Apr
Journey length 0–5,300 miles

ISABELLINE WHEATEAR
Oenanthe isabellina
Wn 10–12 in **Wt** ¾–1¼ oz
Breeds SE Europe & S Asia, Apr to Sep
Winters Tropical & E Africa, Arabia & Pakistan, Oct to Apr
Journey length 300–3,000 miles

PIED WHEATEAR
Oenanthe pleschanka
Wn 9–11 in **Wt** ½–1 oz
Breeds E Europe & W & C Asia, Apr to Sep
Winters E Africa, Oct to Apr
Journey length 1,850–5,600 miles

BLACK-EARED WHEATEAR
Oenanthe hispanica
Wn 9–10 in **Wt** ½–¾ oz
Breeds S Europe & N Africa, Apr to Sep
Winters Tropical Africa, Sep to Mar
Journey length 600–2,175 miles

RED-TAILED WHEATEAR
Oenanthe xanthoprymna
Wn 10–12 in **Wt** ¾–1 oz
Breeds E Turkey, Turkmenistan & Uzbekistan, Apr to Oct
Winters NE Africa, Arabia & Pakistan, Oct to Mar
Journey length 300–1,850 miles

WHITE'S THRUSH
Zoothera dauma
Wn 17–18 in **Wt** 4–7 oz
Breeds NE Asia; partly resident in Himalayas & Indonesia, May to Oct
Winters N India to S China & Philippines, Oct to Apr
Journey length 300–4,000 miles

SONG THRUSH
Turdus philomelos
Wn 12–14 in **Wt** 2½–4 oz
Breeds Europe & NW Asia; resident in W & S Europe, Apr to Oct
Winters W & S Europe, N Africa & Middle East, Oct to Mar
Journey length 0–4,000 miles

MISTLE THRUSH
Turdus viscivorus
Wn 16–18 in **Wt** 3½–6 oz
Breeds Europe & W Asia; resident in W & S Europe, Apr to Oct
Winters W & S Europe, N Africa & SW Asia, Oct to Mar
Journey length 0–3,000 miles

PALLAS' GRASSHOPPER WARBLER
Locustella certhiola
Wn 6–7 in **Wt** ½–¾ oz
Breeds E & C Asia, Jun to Sep
Winters India & Sri Lanka to Indochina & Indonesia, Oct to Apr
Journey length 2,175–3,750 miles

LANCEOLATED WARBLER
Locustella lanceolata
Wn 5–7 in **Wt** ½ oz
Breeds N Asia, Jun to Sep
Winters Indochina & Indonesia, Oct to Apr
Journey length 1,850–5,600 miles

COMMON GRASSHOPPER WARBLER
Locustella naevia
Wn 5–7 in **Wt** ½ oz
Breeds Europe & W Asia, Apr to Sep
Winters W Africa & Indian subcontinent, Oct to Mar
Journey length 1,850–4,000 miles

RIVER WARBLER
Locustella fluviatilis
Wn 7–8 in **Wt** ½–¾ oz
Breeds E Europe & NW Asia, May to Aug
Winters SE Africa, Oct to Apr
Journey length 3,750–5,900 miles

SAVI'S WARBLER
Locustella luscinioides
Wn 7–8 in **Wt** ½–¾ oz
Breeds C Europe & W Asia, Apr to Sep
Winters Tropical Africa, Oct to Mar
Journey length 2,500–3,700 miles

AQUATIC WARBLER
Acrocephalus paludicola
Wn 6–7 in **Wt** ½ oz
Breeds C & E Europe, Apr to Aug
Winters W Africa, Sep to Mar
Journey length 3,000–4,700 miles

SEDGE WARBLER
Acrocephalus schoenobaenus
Wn 6–8 in **Wt** ¼–¾ oz
Breeds Europe except extreme S, NW Asia, Apr to Sep
Winters Tropical & S Africa, Oct to Mar
Journey length 2,500–7,150 miles

PADDYFIELD WARBLER
Acrocephalus agricola
Wn 5–7 in **Wt** ¼–½ oz
Breeds E Europe, W & C Asia, May to Sep
Winters Indian subcontinent, Sep to Apr
Journey length 600–4,000 miles

BLYTH'S REED WARBLER
Acrocephalus dumetorum
Wn 6–7 in **Wt** ¼–½ oz
Breeds E Europe, W & C Asia, May to Aug
Winters Indian subcontinent, Sep to Apr
Journey length 900–4,000 miles

MARSH WARBLER
Acrocephalus palustris
Wn 7–8 in **Wt** ¼–½ oz
Breeds Europe except extreme W & N, May to Sep
Winters E & S Africa, Oct to Apr
Journey length 3,000–6,500 miles

GREAT REED WARBLER
Acrocephalus arundinaceus
Wn 9–11 in **Wt** ¾–1½ oz
Breeds Eurasia except extreme N, Apr to Sep
Winters Tropical & S Africa, Indochina & Indonesia, Oct to Mar
Journey length 1,500–6,500 miles

THICK-BILLED WARBLER
Acrocephalus aedon
Wn 8–9 in **Wt** ¾–1 oz
Breeds S Siberia E to N China, May to Aug
Winters S China, Indochina, India & Burma, Sep to Apr
Journey length 1,850–3,750 miles

BOOTED WARBLER
Hippolais caligata
Wn 7–8 in **Wt** ¼–½ oz
Breeds E Europe & W Asia, May to Sep
Winters Indian subcontinent, Oct to Apr
Journey length 600–4,350 miles

UPCHER'S WARBLER
Hippolais languida
Wn 7–9 in **Wt** ½–¾ oz
Breeds Turkmenistan, Uzbekistan & Iran, Apr to Sep
Winters E Africa, Oct to Mar
Journey length 1,500–3,400 miles

OLIVE-TREE WARBLER
Hippolais olivetorum
Wn 9–10 in **Wt** ½–¾ oz
Breeds Greece & Balkans, May to Aug
Winters SE Africa, Oct to Mar
Journey length 3,000–4,350 miles

ICTERINE WARBLER
Hippolais icterina
Wn 7–9 in **Wt** ½ oz
Breeds Europe except SW, extreme W Asia, May to Aug
Winters S Africa, Oct to Apr
Journey length 3,000–7,500 miles

MELODIOUS WARBLER
Hippolais polyglotta
Wn 6–7 in **Wt** ¼–½ oz
Breeds SW Europe & NW Africa, May to Aug
Winters W Africa, Sep to Apr
Journey length 1,500–3,400 miles

SUBALPINE WARBLER
Sylvia cantillans
Wn 5–7 in **Wt** ¼–½ oz
Breeds SW & S Europe & NW Africa, Apr to Sep
Winters Tropical Africa, Oct to Mar
Journey length 1,500–2,500 miles

MENETRIES' WARBLER
Sylvia mystacea
Wn 5–7 in **Wt** ¼–½ oz
Breeds SW Asia, Mar to Oct
Winters E Africa & Arabia, Oct to Mar
Journey length 600–2,800 miles

RÜPPELL'S WARBLER
Sylvia rueppelli
Wn 7–8 in **Wt** ¼–½ oz
Breeds Greece & Turkey, Apr to Sep
Winters Tropical E Africa, Oct to Mar
Journey length 1,500–2,175 miles

DESERT WARBLER *Sylvia nana*
Wn 5–7 in **Wt** ¼–½ oz
Breeds S Asia; resident in W Sahara, Apr to Oct
Winters NE Africa, Arabia to Pakistan, Oct to Mar
Journey length 300–2,500 miles

ORPHEAN WARBLER
Sylvia hortensis
Wn 7–9 in **Wt** ½–1 oz
Breeds S Europe, N Africa & Iran, Apr to Sep
Winters Tropical Africa, Arabia & Indian subcontinent, Oct to Mar
Journey length 900–2,800 miles

BARRED WARBLER *Sylvia nisoria*
Wn 9–10 in **Wt** ¾–1 oz
Breeds C & E Europe, W Asia, May to Sep
Winters E Africa, Nov to Apr
Journey length 1,850–4,700 miles

LESSER WHITETHROAT
Sylvia curruca
Wn 6–8 in **Wt** ¼–½ oz
Breeds Eurasia except extreme E, May to Aug
Winters Tropical Africa, Arabia & Indian subcontinent, Sep to Mar
Journey length 600–4,000 miles

GARDEN WARBLER *Sylvia borin*
Wn 7–9 in **Wt** ½–1¼ oz
Breeds Europe & NW Asia, Apr to Sep
Winters Tropical & S Africa, Oct to Mar
Journey length 2,500–6,800 miles

BLACKCAP *Sylvia atricapilla*
Wn 7–9 in **Wt** ½–1 oz
Breeds Europe & N Africa; resident in SW Europe & N Africa, Apr to Oct
Winters SW Europe, N, W & E Africa, Nov to Mar
Journey length 0–5,900 miles

GREEN WARBLER
Phylloscopus nitidus
Wn 6–7 in **Wt** ⅕–⅖ oz
Breeds Turkey, Georgia, Iran &

Turkestan, May to Oct
Winters S India, Oct to Apr
Journey length 2,500–3,000 miles

GREENISH WARBLER
Phylloscopus trochiloides
Wn 5–8 in **Wt** ⅕–⅖ oz
Breeds E Europe, W & S Asia, Apr to Sep
Winters India & SE Asia, Oct to Mar
Journey length 600–4,700 miles

PALLAS' LEAF WARBLER
Phylloscopus proregulus
Wn 4–6 in **Wt** ⅕–⅖ oz
Breeds E & S Asia, Apr to Sep
Winters S China, India & Indochina, Oct to Mar
Journey length 600–3,400 miles

YELLOW-BROWED WARBLER
Phylloscopus inornatus
Wn 5–7 in **Wt** ⅕–⅖ oz
Breeds Asia except SE & SW, May to Sep
Winters Arabia, Indian subcontinent, SE Asia & Indochina, Oct to Apr
Journey length 600–5,600 miles

RADDE'S WARBLER
Phylloscopus schwarzi
Wn 6–8 in **Wt** ¼–½ oz
Breeds CE Asia, May to Sep
Winters Burma & Indochina, Oct to Apr
Journey length 1,250–2,500 miles

DUSKY WARBLER
Phylloscopus fuscatus
Wn 5–7 in **Wt** ¼–½ oz
Breeds E Asia, May to Sep
Winters N India E to S China & Indochina, Oct to Apr
Journey length 900–3,000 miles

BONELLI'S WARBLER
Phylloscopus bonelli
Wn 6–7 in **Wt** ⅕–⅖ oz
Breeds S Europe & N Africa, Apr to Aug
Winters Tropical Africa, Sep to Mar
Journey length 1,500–3,000 miles

COMMON CHIFFCHAFF
Phylloscopus collybita
Wn 5–8 in **Wt** ⅕–⅖ oz
Breeds Europe & N Asia except extreme E; resident in SW Europe, Mar to Sep
Winters SW Europe, N & tropical Africa, Arabia & Indian subcontinent, Sep to Mar
Journey length 0–5,600 miles

RED-BREASTED FLYCATCHER
Ficedula parva
Wn 7–8 in **Wt** ¼–½ oz
Breeds NE Europe & N Asia, Apr to Sep
Winters Indian subcontinent E to S China, Oct to Mar
Journey length 1,850–5,000 miles

SEMI-COLLARED FLYCATCHER
Ficedula semitorquata
Wn 9 in **Wt** ½ oz
Breeds SE Europe, Georgia & Turkmenistan, May to Sep
Winters E Africa, Oct to Mar
Journey length 3,000–3,750 miles

COLLARED FLYCATCHER
Ficedula albicollis
Wn 8–9 in **Wt** ¼–½ oz
Breeds E Europe, Apr to Aug
Winters SE Africa, Oct to Mar
Journey length 3,000–6,000 miles

GOLDEN ORIOLE *Oriolus oriolus*
Wn 17–18 in **Wt** 1¾–3 oz
Breeds Europe & W & S Asia, Apr to Aug
Winters S Africa & Indian subcontinent, Sep to Mar
Journey length 600–6,800 miles

LESSER GRAY SHRIKE *Lanius minor*
Wn 12–13 in **Wt** 1½–2 oz
Breeds E Europe & W Asia, Apr to Sep
Winters S Africa, Oct to Mar
Journey length 4,000–6,500 miles

WOODCHAT SHRIKE *Lanius senator*
Wn 10–11 in **Wt** 1–1¾ oz
Breeds S Europe & N Africa, Apr to Sep
Winters Tropical & SE Africa, Oct to Mar
Journey length 1,500–4,000 miles

MASKED SHRIKE *Lanius nubicus*
Wn 9–10 in **Wt** ½–¾ oz
Breeds Greece, Turkey & Middle East, Oct to Mar
Winters Tropical & E Africa, Nov to Feb
Journey length 1,500–2,800 miles

WINTER VISITORS FROM THE FAR NORTH

RED-THROATED LOON
Gavia stellata
Wn 41–45 in
Wt 2 lb 3 oz–4 lb 3 oz
Breeds Circumpolar Arctic & subarctic, May to Sep
Winters Coastal waters of N America & Eurasia, Oct to Apr
Journey length 155–4,700 miles

ARCTIC LOON *Gavia arctica*
Wn 43–51 in
Wt 2 lb 14 oz–7 lb 8 oz
Breeds Circumpolar Arctic & subarctic; also N forested areas with lakes, Apr to Sep
Winters Pacific coasts of N America & Asia & coasts of Europe, Oct to Apr
Journey length 155–4,000 miles

COMMON LOON *Gavia immer*
Wn 50–57 in
Wt 6 lb 3 oz–9 lb 15 oz
Breeds Arctic & subarctic N America, Greenland & Iceland, May to Sep
Winters Coasts of N America & W Europe, Oct to Apr
Journey length 60–4,000 miles

YELLOW-BILLED LOON
Gavia adamsii
Wn 53–59 in
Wt 9 lb 6 oz–14 lb 2 oz
Breeds Circumpolar Arctic except Greenland, May to Sep
Winters N Pacific coasts of N America & Asia & coast of Norway, Oct to Apr
Journey length 300–3,400 miles

PINK-FOOTED GOOSE
Anser brachyrhynchus
Wn 53–66 in
Wt 3 lb 15 oz–7 lb 6 oz
Breeds Iceland, E Greenland & Svalbard, May to Sep
Winters NW Europe, Sep to Apr
Journey length 500–2,500 miles

GREATER WHITE-FRONTED GOOSE
Anser albifrons
Wn 51–64 in
Wt 3 lb 2 oz–7 lb 6 oz
Breeds Circumpolar Arctic, May to Sep
Winters W US, NW & S Europe, Far East, Oct to Apr
Journey length 1,500–3,400 miles

LESSER WHITE-FRONTED GOOSE
Anser erythropus
Wn 47–53 in
Wt 2 lb 14 oz–5 lb 8 oz
Breeds Arctic Eurasia, May to Sep
Winters S Europe, SW Asia, China, Oct to Apr
Journey length 1,850–3,750 miles

ROSS' GOOSE *Anser rossii*
Wn 47–53 in
Wt 2 lb 10 oz–3 lb 15 oz
Breeds Arctic Canada, May to Sep
Winters W US, Oct to Apr
Journey length 2,300–2,900 miles

GREATER SCAUP *Aythya marila*
Wn 28–33 in **Wt** 1 lb 10 oz–3 lb
Breeds Circumpolar Arctic, May to Sep
Winters Coasts of N America & Eurasia, Oct to Apr
Journey length 300–4,000 miles

KING EIDER *Somateria spectabilis*
Wn 33–40 in
Wt 2 lb 11 oz–4 lb 7 oz
Breeds Circumpolar Arctic, Jun to Sep
Winters Coasts of E US, Alaska, E Siberia & Norway, Oct to May
Journey length 600–3,400 miles

SPECTACLED EIDER
Somateria fischeri
Wn 33–36 in
Wt 3 lb 1 oz–4 lb 1 oz
Breeds Coasts of NE Siberia & N Alaska, Jun to Sep
Winters Bering Sea, Oct to May
Journey length 185–500 miles

STELLER'S EIDER *Polysticta stelleri*
Wn 27–29 in
Wt 1 lb 2 oz–2 lb 3 oz
Breeds Coasts of E Siberia & Alaska, Jun to Sep
Winters Coasts of S Alaska & NE Asia, Oct to May
Journey length 300–900 miles

BLACK SCOTER *Melanitta nigra*
Wn 31–35 in
Wt 1 lb 9 oz–3 lb 3 oz
Breeds Arctic Eurasia & Alaska, May to Sep
Winters Coasts of N America, Eurasia & NW Africa, Oct to Apr
Journey length 300–4,350 miles

WHITE-WINGED SCOTER
Melanitta fusca
Wn 35–38 in **Wt** 3 lb–4 lb 6 oz
Breeds Circumpolar Arctic & subarctic, May to Sep
Winters Coasts of N America & Eurasia, Oct to Apr
Journey length 125–3,000 miles

SURF SCOTER
Melanitta perspicillata
Wn 30–36 in
Wt 1 lb 7 oz–2 lb 8 oz
Breeds Arctic N America, May to Sep
Winters Coasts of N America, Oct to Apr
Journey length 300–2,800 miles

GYRFALCON *Falco rusticolus*
Wn 51–62 in
Wt 1 lb 12 oz–4 lb 10 oz
Breeds Circumpolar Arctic, May to Sep
Winters N Canada & Eurasia, Oct to Apr
Journey length 300–1,500 miles

EURASIAN DOTTEREL
Charadrius morinellus
Wn 22–25 in **Wt** 3–4 oz
Breeds Arctic Eurasia, also mountain tops farther S, May to Oct
Winters N Africa & Middle East, Nov to Apr
Journey length 1,850–5,600 miles

EUROPEAN GOLDEN PLOVER
Pluvialis apricaria
Wn 26–29 in **Wt** 5–10 oz
Breeds Arctic & subarctic Europe & W Asia, Apr to Sep
Winters W & S Europe & N Africa, Oct to Mar
Journey length 300–4,000 miles

BLACK-BELLIED PLOVER
Pluvialis squatarola
Wn 27–32 in **Wt** 6–12 oz
Breeds Circumpolar Arctic, May to Sep
Winters Coasts of S N America, S America, S Africa, Arabia, Indian subcontinent, SE Asia & Australasia, Oct to Apr
Journey length 1,500–8,700 miles

GREAT KNOT *Calidris tenuirostris*
Wn 24–25 in **Wt** 5–7 oz
Breeds NE Siberia, May to Sep
Winters Pakistan to S China, S to Australia, Oct to Apr
Journey length 3,000–8,700 miles

SANDERLING *Calidris alba*
Wn 15–17 in **Wt** 1–2½ oz
Breeds Arctic Canada, Greenland & NW Siberia, May to Aug
Winters Coasts of N & S America, Africa, Arabia, Indian subcontinent, SE Asia & Australasia, Sep to Apr
Journey length 1,500–9,600 miles

SEMIPALMATED SANDPIPER
Calidris pusilla
Wn 13–14 in **Wt** ¾–1½ oz
Breeds Arctic N America, Jun to Sep
Winters Coasts of C & S America S to Peru & Uruguay, Oct to May
Journey length 4,350–7,150 miles

WESTERN SANDPIPER
Calidris mauri
Wn 13–14 in **Wt** ½–1¼ oz
Breeds NE Siberia & Alaska, May to Sep
Winters S US, W Indies, C America & S America S to Peru, Oct to Apr
Journey length 4,700–6,800 miles

RED-NECKED STINT
Calidris ruficollis
Wn 13–14 in **Wt** ¾–1½ oz
Breeds NE Siberia, Jun to Sep
Winters S China S to Australia & NZ, Oct to Apr
Journey length 3,000–8,700 miles

LITTLE STINT *Calidris minuta*
Wn 13–14 in **Wt** ½–1½ oz
Breeds Arctic Eurasia except NE Siberia, Jun to Aug
Winters S Europe, tropical & S Africa, Arabia & Indian subcontinent, Sep to Apr
Journey length 2,175–8,400 miles

TEMMINCK'S STINT
Calidris temminckii
Wn 13–14 in **Wt** ½–1¼ oz
Breeds Arctic Eurasia, May to Aug
Winters S Europe, tropical Africa, SE Asia & Indian subcontinent, Sep to Apr
Journey length 1,250–5,300 miles

LEAST SANDPIPER
Calidris minutilla
Wn 12–13 in **Wt** ¾–1¼ oz
Breeds Alaska & N Canada, May to Aug
Winters Coasts of S US, C America, W Indies & N S America, Sep to Apr
Journey length 1,500–5,300 miles

WHITE-RUMPED SANDPIPER
Calidris fuscicollis
Wn 15–17 in **Wt** ¾–1½ oz
Breeds Arctic Canada & Alaska, Jun to Aug
Winters S America from equator to Tierra del Fuego, Sep to Apr
Journey length 5,300–9,300 miles

BAIRD'S SANDPIPER
Calidris bairdii
Wn 15–18 in **Wt** 1–1½ oz
Breeds NE Siberia, Alaska, Arctic Canada & NW Greenland, Jun to Aug
Winters S America, Sep to Apr
Journey length 5,000–9,600 miles

PECTORAL SANDPIPER
Calidris melanotos
Wn 16–19 in **Wt** 1½–3¾ oz
Breeds NE Siberia, Alaska, Arctic Canada, May to Aug
Winters S America, Australia & NZ, Sep to Apr
Journey length 5,600–9,600 miles

SHARP-TAILED SANDPIPER
Calidris acuminata
Wn 16–18 in **Wt** 1½–4 oz
Breeds N Siberia, May to Aug
Winters Pacific islands S of equator S to Australia & NZ, Sep to Apr
Journey length 5,600–9,000 miles

CURLEW SANDPIPER
Calidris ferruginea
Wn 16–18 in **Wt** 1¼–3½ oz
Breeds N Siberia, Jun to Aug
Winters Tropical & S Africa, coasts of Indian subcontinent, SE Asia & Australasia, Sep to May
Journey length 3,750–9,000 miles

ROCK SANDPIPER
Calidris ptilocnemis
Wn 16–18 in **Wt** 2–4 oz
Breeds Alaska, May to Sep
Winters W coasts of Canada & US, Oct to Apr
Journey length 300–1,850 miles

PURPLE SANDPIPER
Calidris maritima
Wn 16–18 in **Wt** 2–4 oz
Breeds Arctic Europe, NW Siberia, Iceland, Greenland & NE Canada, May to Aug
Winters Coasts of NW Europe, Iceland & E US, Sep to Apr
Journey length 125–3,750 miles

STILT SANDPIPER
Micropalama himantopus
Wn 16–18 in **Wt** 1½–2¼ oz
Breeds Arctic & subarctic N America, May to Aug
Winters C S America, Sep to Apr
Journey length 5,300–6,800 miles

BUFF-BREASTED SANDPIPER
Tryngites subruficollis
Wn 16–18 in **Wt** 1¾–3¼ oz
Breeds Alaska & NW Arctic Canada, May to Aug
Winters C Argentina & Paraguay, Sep to Apr
Journey length 5,600–6,800 miles

LONG-BILLED DOWITCHER
Limnodromus scolopaceus
Wn 18–20 in **Wt** 3¼–5 oz
Breeds NE Siberia & Alaska, May to Sep
Winters S US S to Guatemala, Oct to Apr
Journey length 3,000–5,000 miles

BAR-TAILED GODWIT
Limosa lapponica
Wn 27–31 in **Wt** 7 oz–1 lb
Breeds Arctic Eurasia & NW Alaska, May to Sep
Winters Coasts of W Europe, Africa, Madagascar, Arabia, Indian subcontinent, SE Asia & Australasia, Oct to Apr
Journey length 1,250–9,000 miles

HUDSONIAN GODWIT
Limosa haemastica
Wn 25–29 in **Wt** 6–14 oz
Breeds Alaska, NE Arctic Canada & Hudson Bay, May to Sep
Winters S America, Oct to Apr
Journey length 5,600–9,300 miles

BRISTLE-THIGHED CURLEW
Numenius tahitiensis
Wn 28–32 in **Wt** 7–16 oz
Breeds W Alaska, Jun to Sep
Winters Pacific islands, Oct to Apr
Journey length 5,600–6,000 miles

WANDERING TATTLER
Heteroscelus incanus
Wn 16–20 in **Wt** 3–5 oz
Breeds Alaska & NW Canada, May to Aug
Winters Mexico, S America & Australasia, Sep to Apr
Journey length 2,500–9,300 miles

RUDDY TURNSTONE
Arenaria interpres
Wn 19–22 in **Wt** 3–7 oz
Breeds NW Europe & circumpolar Arctic, May to Aug
Winters Coasts of N & S America, Europe, Africa, Arabia, Indian subcontinent, SE Asia & Australasia, Sep to Apr
Journey length 300–9,300 miles

BLACK TURNSTONE
Arenaria melanocephala
Wn 18–21 in **Wt** 2–6 oz
Breeds Alaska, May to Sep
Winters W coasts of Canada, US & Mexico, Oct to Apr
Journey length 600–3,750 miles

SURFBIRD *Aphriza virgata*
Wn 21–23 in **Wt** 3–6 oz
Breeds Alaska, May to Sep
Winters W coasts US, C & S America, Oct to Apr
Journey length 1,250–9,300 miles

RED-NECKED PHALAROPE
Phalaropus lobatus
Wn 12–16 in **Wt** 1–1¾ oz
Breeds Circumpolar Arctic, May to Aug
Winters Offshore from W S America, Arabian peninsula & Indonesia, Sep to Apr
Journey length 3,000–6,500 miles

RED PHALAROPE
Phalaropus fulicarius
Wn 15–17 in **Wt** 1¼–2¾ oz
Breeds Circumpolar Arctic, May to Aug
Winters Offshore W S America, W & SW Africa, Sep to Apr
Journey length 2,500–9,300 miles

POMARINE JAEGER
Stercorarius pomarinus
Wn 49–54 in **Wt** 1 lb 3 oz–2 lb
Breeds Circumpolar Arctic, May to Sep
Winters Offshore in N & W S Atlantic, W, S & E N Pacific and Indian oceans, Oct to Apr
Journey length 600–9,300 miles

PARASITIC JAEGER
Stercorarius parasiticus
Wn 43–49 in **Wt** 11 oz–1 lb 7 oz
Breeds Circumpolar Arctic & subarctic, May to Sep
Winters Offshore in N & S Atlantic, W, S & E N Pacific and Indian oceans, Oct to Apr
Journey length 600–10,000 miles

SABINE'S GULL *Larus sabini*
Wn 35–39 in **Wt** 5–8 oz
Breeds Arctic Canada, Alaska, N & E Siberia, Jun to Aug
Winters Offshore E coasts of Atlantic & Pacific, Sep to May
Journey length 1,250–8,700 miles

ICELAND GULL *Larus glaucoides*
Wn 55–59 in
Wt 1 lb 10 oz–1 lb 14 oz
Breeds Greenland & Baffin Island, May to Aug
Winters N Atlantic, Sep to Apr
Journey length 600–1,850 miles

GLAUCOUS GULL
Larus hyperboreus
Wn 59–64 in **Wt** 2 lb 6 oz–4 lb
Breeds Circumpolar Arctic, Apr to Oct
Winters N Atlantic & N Pacific, Nov to Mar
Journey length 300–1,850 miles

THAYER'S GULL *Larus thayeri*
Wn 55–59 in **Wt** 1 lb 10 oz–2 lb
Breeds C Arctic Canada, May to Aug
Winters Coasts of SW Canada & W US, Sep to Apr
Journey length 1,500–3,000 miles

GLAUCOUS-WINGED GULL
Larus glaucescens
Wn 55–59 in **Wt** 2 lb–3 lb 8 oz
Breeds Alaska & NW Canada, Apr to Aug
Winters Coasts of W US & NW Mexico, Sep to Apr
Journey length 300–2,800 miles

ROSS' GULL *Rhodostethia rosea*
Wn 35–39 in **Wt** 4–9 oz
Breeds NE Siberia, Jun to Aug
Winters N Pacific, Sep to May
Journey length 1,500–3,000 miles

ALEUTIAN TERN *Sterna aleutica*
Wn 29–31 in **Wt** 3½–4 oz
Breeds E Siberia & W Alaska, May to Sep
Winters N Pacific, Oct to Apr
Journey length 600–1,850 miles

THICK-BILLED MURRE *Uria lomvia*
Wn 25–28 in
Wt 1 lb 10 oz–2 lb 11 oz
Breeds Cirumpolar Arctic coasts & islands, May to Aug
Winters N Pacific & N Atlantic, Sep to Apr
Journey length 900–1,850 miles

BLACK GUILLEMOT *Cepphus grylle*
Wn 20–22 in **Wt** 10 oz–1 lb 2 oz
Breeds Arctic & N coasts & islands, Apr to Aug

Winters N Atlantic & Arctic oceans, Sep to Mar
Journey length 125–1,500 miles

DOVEKIE *Alle alle*
Wn 15–18 in **Wt** 4–6 oz
Breeds Greenland, Iceland, Svalbard & islands N of Siberia, May to Aug
Winters N Atlantic, Sep to Apr
Journey length 300–3,000 miles

CRESTED AUKLET
Aethia cristatella
Wn 15–18 in **Wt** 7–10 oz
Breeds NE Siberia & Alaska, Apr to Aug
Winters Bering Sea & N Pacific, Sep to Mar
Journey length 300–1,250 miles

HORNED LARK
Eremophila alpestris
Wn 11–13 in **Wt** 1–1¾ oz
Breeds Circumpolar Arctic; resident in US, SE Europe & C Asia, May to Sep
Winters Coasts of NW Europe & N America, Oct to Apr
Journey length 600–3,000 miles

RED-THROATED PIPIT
Anthus cervinus
Wn 9–10 in **Wt** ½–¾ oz
Breeds Arctic Eurasia, May to Sep
Winters Tropical & E Africa, SE Asia & Indonesia, Oct to Apr
Journey length 3,750–5,900 miles

SIBERIAN THRUSH *Zoothera sibirica*
Wn 13–14 in **Wt** 2–2¾ oz
Breeds Arctic & N Siberia, May to Sep
Winters India, SE Asia & Indonesia, Oct to Apr
Journey length 2,500–4,700 miles

VARIED THRUSH *Ixoreus naevius*
Wn 14–15 in **Wt** 2¼–3 oz
Breeds Alaska & NW Canada; resident in S of range, Apr to Sep
Winters British Columbia, W US & NW Mexico, Nov to Mar
Journey length 0–3,750 miles

GRAY-CHEEKED THRUSH
Catharus minimus
Wn 11–12 in **Wt** ¾–1½ oz
Breeds Alaska & N Canada, Jun to Sep
Winters N S America, Oct to Apr
Journey length 3,000–6,000 miles

ARCTIC WARBLER
Phylloscopus borealis
Wn 6–8 in **Wt** ¼–½ oz
Breeds Arctic & subarctic Eurasia, Jun to Aug
Winters SE Asia & Indonesia, Sep to Apr
Journey length 1,500–5,900 miles

HOARY REDPOLL
Acanthis hornemanni
Wn 6–8 in **Wt** ½–¾ oz
Breeds Circumpolar Arctic, Apr to Sep
Winters N America & N Eurasia, Oct to Apr
Journey length 300–1,250 miles

HARRIS' SPARROW
Zonotrichia querula
Wn 7–10 in **Wt** ½–1 oz
Breeds C Arctic Canada, May to Aug
Winters Mississippi valley, Sep to Apr
Journey length 1,850–2,800 miles

SMITH'S LONGSPUR
Calcarius pictus
Wn 6–8 in **Wt** ½–1 oz
Breeds Arctic Canada, May to Aug
Winters Mississippi valley, Sep to Apr
Journey length 1,850–2,800 miles

SOUTH AMERICAN MIGRANTS

GREAT GREBE *Podiceps major*
Wn 39–42 in
Wt 3 lb 5 oz–3 lb 12 oz
Breeds NW Peru, Paraguay & SE Brazil S to Patagonia & S Chile, Oct to Mar, though can breed year-round
Winters Coastal waters of S America, Apr to Sep
Journey length 125–900 miles

BLACK-FACED IBIS
Theristicus melanopis
Wn 39–43 in
Wt 2 lb 7 oz–3 lb 1 oz
Breeds S Chile & S Argentina; resident in Ecuador, Peru & NW Bolivia, Sep to Mar
Winters N Argentina, Apr to Sep
Journey length 300–2,175 miles

BLACK-NECKED SWAN
Cygnus melanocorypha
Wn 55–66 in
Wt 7 lb 11 oz–14 lb 12 oz
Breeds Tierra del Fuego N to C Chile, where sedentary, Jul to Jan/Feb
Winters Argentina & Chile, except S, Paraguay & S Brazil, Feb to Jul
Journey length 125–2,175 miles

RED SHOVELER *Anas platalea*
Wn 27–33 in **Wt** 1 lb 2 oz–1 lb 5 oz
Breeds Tierra del Fuego N to C Chile & N Argentina, where sedentary, Sep to Feb
Winters C Argentina & C Chile N to S Brazil & S Peru, Feb to Aug
Journey length 300–3,000 miles

ARGENTINE BLUE-BILLED DUCK
Oxyura vittata
Wn 21–24 in **Wt** 1 lb 4 oz–1 lb 6 oz
Breeds Chile & Argentina N to SE Brazil, where sedentary, Oct to Feb
Winters C Chile & C Argentina N to Paraguay & C Brazil, Mar to Sep
Journey length 125–2,175 miles

PLUMBEOUS KITE
Ictinia plumbea
Wn 33–37 in **Wt** 7–10 oz
Breeds N Argentina & Paraguay N to C Mexico, Oct to Apr in S, Mar to Sep in N
Winters Argentinian breeders to Brazil, Apr to Sep; Mexican & C American breeders to N S America, Sep to Mar
Journey length 300–2,500 miles

LONG-WINGED HARRIER
Circus buffoni
Wn 45–51 in **Wt** 14 oz–1 lb 7 oz
Breeds Argentina, except S, Bolivia & Chile; resident in Colombia & Trinidad, Oct to Mar
Winters N S America, Mar to Sep
Journey length 300–3,000 miles

LEAST SEEDSNIPE
Thinocorus rumicivorus
Wn 14–15 in **Wt** 2–3 oz
Breeds Argentina & Chile N to Ecuador; resident in N of range, Oct to Feb
Winters N Chile & Bolivia to Colombia, Mar to Sep
Journey length 300–3,750 miles

RUFOUS-CHESTED DOTTEREL
Onaradrius modestus
Wn 22–25 in **Wt** 3–4 oz
Breeds Tierra del Fuego & S Chile, Oct to Feb
Winters Argentina, Chile, Uruguay & E Brazil, Mar to Sep
Journey length 125–2,000 miles

GREEN-BACKED FIRECROWN
Sephanoides sephanoides
Wn 5–6 in **Wt** 1/15 oz
Breeds Tierra del Fuego N to C Chile & W Argentina, Oct to Feb
Winters E Argentina, Mar to Sep
Journey length 125–900 miles

STRIPED CUCKOO *Tapera naevia*
Wn 19–21 in **Wt** 2 oz
Breeds S Mexico S to N Argentina & SE Brazil, Mar to Jun in N; Oct to Jan in S
Winters S breeders move N, Mar to Sep
Journey length 300–900 miles

FORK-TAILED FLYCATCHER
Tyrannus savana
Wn 9–11 in **Wt** 1 oz
Breeds C Argentina & Uruguay N to SE Mexico; sedentary around equator, Oct to Jan in S, Mar to Jun in N
Winters N pops. to N S America, Jul to Feb; S pops. to N Argentina & Brazil, Feb to Sep
Journey length 300–1,250 miles

LARGE ELAENIA *Elaenia spectabilis*
Wn 10–11 in **Wt** 1 oz
Breeds S Brazil & N Argentina, Oct to Jan
Winters Amazonia, Feb to Sep
Journey length 300–1,500 miles

BROWN-CHESTED MARTIN
Progne tapera
Wn 13–15 in **Wt** ¾–1 oz
Breeds N Argentina & S Brazil N to N Colombia & Guyana, also SW Ecuador & W Peru; sedentary in N, Oct to Feb
Winters Bolivia & S Brazil N to C Panama, Mar to Sep
Journey length 300–1,500 miles

AFRICAN MIGRANTS

DWARF BITTERN
Ixobrychus sturmii
Wn 17–19 in **Wt** 3–4 oz
Breeds Tropics S from Senegal, Nigeria, Chad & S Sudan, Jun to Sep
Winters S from breeding range to E Cape Province & Transvaal, Nov to Apr
Journey length 300–3,000 miles

MALAGASY POND HERON
Ardeola idae
Wn 31–35 in **Wt** 8 oz
Breeds Madagascar & Aldabra, Nov to May
Winters C & E Africa, May to Oct
Journey length 600–1,850 miles

ABDIM'S STORK
Ciconia abdimii
Wn 43–51 in **Wt** 2 lb 10 oz–3 lb 1 oz
Breeds Tropics from Senegal to Ethiopia & extreme SW Arabia, Apr to Sep
Winters E & S Africa, Nov to Mar
Journey length 600–4,350 miles

SOUTHERN POCHARD
Aythya erythrophthalma
Wn 29–33 in
Wt 1 lb 3 oz–2 lb 3 oz
Breeds Ethiopia S to Cape Province; mainly sedentary, S breeders disperse N in dry season, timing variable
Winters S birds move N to Kenya & Uganda, Sep to Mar
Journey length 1,250–3,000 miles

AFRICAN SCISSOR-TAILED KITE
Chelictinia riocourii
Wn 25–29 in **Wt** 6–7 oz
Breeds Tropics from Senegal to Ethiopia, Somalia & N Kenya, Mar to Aug
Winters Gambia & N Nigeria in W, C Kenya in E, Oct to Feb
Journey length 125–900 miles

WAHLBERG'S EAGLE
Aquila wahlbergi
Wn 49–55 in **Wt** 15 oz–1 lb 14 oz
Breeds Savanna S from Gambia, Niger, Chad & Eritrea to N Cape Province, Sep/Oct to Mar/Apr
Winters Mainly N, but some sedentary, May to Aug
Journey length 60–600 miles

MADAGASCAR PRATINCOLE
Glareola ocularis
Wn 23–26 in **Wt** 3–4 oz
Breeds E Madagascar, Sep to Mar
Winters Coasts of E Africa from N Mozambique N to S Somalia, Apr to Sep
Journey length 500–1,100 miles

BROWN-CHESTED PLOVER
Vanellus superciliosus
Wn 23–25 in **Wt** 3½–4 oz
Breeds Ghana E to Central African Republic, Jan to May
Winters Zaïre, Tanzania & Uganda, Jun to Dec
Journey length 600–2,500 miles

DAMARA TERN *Sterna balaenarum*
Wn 19–20 in **Wt** 1–1½ oz
Breeds W coast of Africa from Angola to S Africa, Nov to May
Winters N Gulf of Guinea, Jul to Oct
Journey length 1,250–2,800 miles

JACOBIN CUCKOO
Clamator jacobinus
Wn 17–19 in **Wt** 2–3 oz
Breeds Sub-Saharan Africa from Senegal to Red Sea, S to Cape Province; other races Indian subcontinent, Oct to Mar in S, May to Jul in N
Winters N movement of N breeders to Ethiopia, Sudan & Chad, Oct to Apr; N movement of S breeders up to & perhaps N of equator, Apr to Sep
Journey length 600–2,175 miles

AFRICAN CUCKOO *Cuculus gularis*
Wn 21–23 in **Wt** 3–5 oz
Breeds Senegal to Ethiopia & S to S Africa, Oct to Mar in S, Apr to Jul in N
Winters N of breeding range, timing variable but generally May to Sep in S, Sep to Mar in N
Journey length 300–1,250 miles

AFRICAN EMERALD CUCKOO
Chrysococcyx cupreus
Wn 13–15 in **Wt** 1¼–1½ oz
Breeds Sub-Saharan Africa from Gambia, Sudan & Ethiopia S to S Africa; timing variable to occur with that of weavers; Oct to Mar in S, May to Jul in N
Winters N of breeding range, Apr to Sep in S, Aug to Apr in N
Journey length 300–600 miles

DIDRIC CUCKOO
Chrysococcyx caprius
Wn 12–13 in **Wt** ½–1½ oz
Breeds Tropics from Senegal to Ethiopia & S to S Africa, timing variable, Sep to Apr in S, Jun to Sep in N
Winters S breeders move N perhaps as far as equator, May to Aug; N breeders move N, Oct to May; some resident in equatorial region, where timing variable
Journey length 300–1,850 miles

PENNANT-WINGED NIGHTJAR
Macrodipteryx vexillarius
Wn 25–26 in **Wt** 2–3 oz
Breeds Tropics S of equator to S Tanzania, Zambia & Angola, Sep to Jan/Feb
Winters S Sudan W to Nigeria, Feb/Mar to Sep
Journey length 1,250–2,500 miles

WHITE-RUMPED SWIFT *Apus caffer*
Wn 13–14 in **Wt** ¾–1 oz
Breeds Tropical W & E Africa S to S Africa, Sep to Mar; small numbers Spain & Morocco, Mar to Jun
Winters S breeders to equator, Apr to Aug; W African breeders may be resident, may disperse N, Nov to Jan
Journey length 300–2,500 miles

GRAY-HEADED KINGFISHER
Halcyon leucocephala
Wn 12–13 in **Wt** 1¼–1¾ oz
Breeds Senegal to Ethiopia & N Somalia S to N Zaïre, Kenya & Tanzania; resident in Cape Verde Islands, Mar to May in N, Dec to Mar in S
Winters C Nigerian breeders to S Nigeria Nov to Feb before breeding, to Sudan & Niger after, May to Oct
Journey length 600–1,500 miles

BLUE-CHEEKED BEE-EATER
Merops superciliosus
Wn 18–19 in **Wt** 1¼–2 oz
Breeds Tropical W Africa, E, NW & S Africa; also SW Asia & India; Mar to Jul in W, May to Sep in N, Oct to Mar in S
Winters S African breeders N to Zaïre & E Africa, May to Sep; W African breeders resident; N African breeders to W Africa E to Nigeria, Oct to Apr
Journey length 300–5,000 miles

RED-BREASTED SWALLOW
Hirundo semirufa
Wn 12–13 in **Wt** ½–¾ oz
Breeds Tropics from Senegal to Ethiopia S to S Africa, Apr to Jul in N, Nov to Feb in S
Winters Some N movement though more sedentary in N, Aug to Feb in N; Mar to Aug in S
Journey length 300–900 miles

AFRICAN REED WARBLER
Acrocephalus baeticatus
Wn 6–7 in **Wt** ¼–½ oz
Breeds Tropics from Senegal to Somalia S to Angola, Mozambique & Zambia, Apr to Jul in N, Nov to Feb in S
Winters N; S breeders travel farther, Aug to Feb in N, Mar to Aug in S
Journey length 300–1,250 miles

AMETHYST STARLING
Cinnyricinclus leucogaster
Wn 9–10 in **Wt** 1–1¼ oz
Breeds Tropics from Senegal to Somalia & S to Namibia & Botswana; also SW Arabia, Mar to Jul in N, Oct to Feb in S
Winters Toward equator; some sedentary or nomadic, Aug to Feb in N, Mar to Aug in S
Journey length 300–900 miles

AUSTRALASIAN MIGRANTS

CATTLE EGRET *Bubulcus ibis*
Wn 34–37 in **Wt** 12–14 oz
Breeds S & E Asia to Australia & NZ; also Africa, S Europe, S, C & S N America, Sep to Feb
Winters Australian breeders winter SE Australia, Tasmania & NZ; some from NW Australia winter in SW, Mar to Aug
Journey length 300–2,800 miles

GRAY FALCON *Falco hypoleucos*
Wn 37–39 in **Wt** 1 lb–1 lb 9 oz
Breeds Inland Australia to coast in W; Jul to Dec, Nov to Mar in W
Winters N Australia & New Guinea, Jan to Jun
Journey length 300–1,250 miles

AUSTRALIAN HOBBY
Falco longipennis
Wn 32–36 in **Wt** 5–11 oz
Breeds Australia except C desert, Sep to Mar
Winters E Indonesia & New Guinea, Apr to Aug
Journey length 500–900 miles

SOUTH ISLAND PIED OYSTERCATCHER
Haematopus finschi
Wn 31–33 in **Wt** 1 lb–1 lb 3 oz
Breeds C & E parts of S Island, NZ, Sep to Feb
Winters N & E coasts of S Island, coasts of N Island & Stewart Island, NZ, Feb to Aug
Journey length 60–750 miles

WRYBILL *Anarhynchus frontalis*
Wn 20–22 in **Wt** 2–2½ oz
Breeds C S Island, NZ, Jul to Dec
Winters N N Island, NZ, Dec to Jul
Journey length 500–870 miles

AUSTRALIAN PRATINCOLE
Stiltia isabella
Wn 21–23 in **Wt** 2¼–2¾ oz
Breeds E, inland & N Australia, probably resident in N, timing variable, May to Dec
Winters N Australia N to New Guinea, Java, Borneo & Sulawesi, Dec to Jun
Journey length 300–2,175 miles

DOUBLE-BANDED PLOVER
Charadrius bicinctus
Wn 17–21 in **Wt** 2–3 oz
Breeds N, S and Chatham Islands, NZ; sedentary on Auckland Islands, Sep to Feb
Winters N N Island, NZ, & SE & E Australia, Mar to Aug
Journey length 300–1,500 miles

ORANGE-BELLIED PARROT
Neophema chrysogaster
Wn 11–13 in **Wt** 1¾–2 oz
Breeds Tasmania, Nov to Mar
Winters Coastal SE Australia, Victoria & Tasmania, Apr to Oct
Journey length 155–470 miles

COCKATIEL *Nymphicus hollandicus*
Wn 14–17 in **Wt** 2–3 oz
Breeds Australia, except coasts, timing variable, Aug to Mar
Winters Southerly breeders to N, Apr to Aug, also nomadic
Journey length 300–600 miles

CHANNEL-BILLED CUCKOO
Scythrops novaehollandiae
Wn 51–62 in
Wt 1 lb 5 oz–1 lb 12 oz
Breeds N tropical Australia S to New S Wales & N S Australia, Oct to Jan
Winters N Australia & New Guinea to Sulawesi, Feb/Mar to Sep
Journey length 300–2,175 miles

LONG-TAILED CUCKOO
Urodynamis taitensis
Wn 17–20 in **Wt** 2–2¼ oz
Breeds NZ, Sep to Feb
Winters Samoa, Tonga, Fiji, Mar to Aug
Journey length 1,500–2,175 miles

WHITE-THROATED NIGHTJAR
Eurostopodus mystacalis
Wn 24–25 in **Wt** 2–2½ oz
Breeds E Australia, Queensland to Victoria, Oct to Jan
Winters N Australia & New Guinea, Feb to Sep
Journey length 300–900 miles

BUFF-BREASTED PARADISE KINGFISHER *Tanysiptera sylvia*
Wn 7–9 in **Wt** ¾ oz
Breeds N Queensland; resident in New Guinea, Nov to Jan
Winters New Guinea, Feb to Oct
Journey length 300–900 miles

DOLLARBIRD
Eurystomus orientalis
Wn 21–23 in **Wt** 4–5 oz
Breeds E India to Japan, China, Philippines, S to Australia & NZ, Apr to Aug in N, Oct to Jan in S
Winters N breeders S to Malaysia, Borneo & Sumatra, Sep to Mar; S breeders N to New Guinea & E Indonesia, Feb to Sep
Journey length 600–1,850 miles

RED-BREASTED PITTA
Pitta erythrogaster
Wn 15–17 in **Wt** 2–3 oz
Breeds NE Queensland, Oct to Jan
Winters New Guinea, Indonesia & Philippines, Feb to Sep
Journey length 600–2,500 miles

TREE MARTIN *Hirundo nigricans*
Wn 10–11 in **Wt** ½–¾ oz
Breeds S Australia, Aug to Jan
Winters N Australia N to New Caledonia, New Guinea & Indonesia, Feb to Jul
Journey length 600–3,000 miles

COMMON CICADABIRD
Coracina tenvirostris
Wn 12–14 in **Wt** 1½–1¾ oz
Breeds E & SE Australia, Oct to Jan
Winters N & E Australia, Feb to Sep
Journey length 300–1,250 miles

WHITE-WINGED TRILLER
Lalage sucurii
Wn 10–11 in **Wt** 1–1¼ oz
Breeds Throughout Australia except C desert, Sep to Jan/Feb
Winters New Guinea & Indonesia, Feb/Mar to Aug
Journey length 600–2,175 miles

FLAME ROBIN *Petroica phoenicea*
Wn 8–9 in **Wt** ½ oz

Breeds E Australia from SE Queensland S to Victoria & Tasmania, Sep to Jan
Winters At lower altitudes & W into drier regions, Feb to Aug
Journey length 60–300 miles

AUSTRALIAN REED WARBLER
Acrocephalus australis
Wn 8–9 in **Wt** ¾–1 oz
Breeds S Australia, except drier areas, & Tasmania, Sep to Jan
Winters N & NE Australia, Feb to Aug
Journey length 300–900 miles

RUFOUS SONGLARK
Cinclorhamphus mathewsi
Wn 7–9 in **Wt** ¾–1 oz
Breeds Throughout Australia except extreme N; resident in N of range, Aug to Feb
Winters N & NE Australia, Mar to Jul
Journey length 300–900 miles

WHITE-THROATED GERYGONE
Gerygone olivacea
Wn 6–8 in **Wt** ¼–½ oz
Breeds N, E & SE Australia, Aug to Jan
Winters N Australia & New Guinea, Feb to Jul/Aug
Journey length 125–900 miles

BLACK-FACED MONARCH
Monarcha melanopsis
Wn 9–11 in **Wt** 1–1¼ oz
Breeds E Australia from NE Queensland to Victoria; probably resident in N, Nov to Jan
Winters N Australia & New Guinea, Feb to Oct
Journey length 300–1,250 miles

SATIN FLYCATCHER
Myiagra cyanoleuca
Wn 9–11 in **Wt** 1–1¼ oz
Breeds E & SE Australia & Tasmania; probably resident in N, Nov to Jan
Winters E Australia, Feb to Oct
Journey length 125–1,100 miles

RUFOUS FANTAIL
Rhipidura rufifrons
Wn 8–9 in **Wt** ½–¾ oz
Breeds New Guinea, Pacific & Indonesian islands, N, E & SE Australia & Tasmania; resident in N, Nov to Jan
Winters New Guinea, Pacific & Indonesian islands, N & E Australia, Feb to Oct
Journey length 300–900 miles

YELLOW-FACED HONEYEATER
Meliphaga chrysops
Wn 7–8 in **Wt** ½ oz
Breeds E & SE Australia & Tasmania; resident in N of range, Jul to Jan
Winters E Australia, Feb to Jun
Journey length 125–900 miles

EASTERN SPINEBILL
Acanthorhynchus tenuirostris
Wn 6–7 in **Wt** ½ oz
Breeds E, SE Australia & Tasmania; resident in N of range, Aug to Jan
Winters E Australia, Feb to Jul
Journey length 125–900 miles

MIGRATORY SEA BIRDS

KING PENGUIN
Aptenodytes patagonicus
Wt 19 lb 13 oz–33 lb 1 oz
Breeds Falklands, S Georgia, islands in S Ocean, mainly Sep, lasting 15 months; breeds twice every 3 yrs
Winters At sea, Sep to Sep
Journey length 250–600 miles

GENTOO PENGUIN
Pygoscelis papua
Wt 4 lb 7 oz–6 lb 10 oz
Breeds Antarctic Peninsula & islands in S Ocean; sedentary around latter, Jun/Nov (varies with colony) to Apr
Winters At sea, Apr to Jun/Nov
Journey length 300–1,500 miles

LITTLE PENGUIN *Eudyptula minor*
Wt 2 lb–2 lb 7 oz
Breeds S Australia, NZ & neighboring islands; timing variable, breeding recorded in all months
Winters At sea: mainly sedentary, timing variable
Journey length 0–155 miles

GALÁPAGOS PENGUIN
Spheniscus mendiculus
Wt 2 lb 3 oz–3 lb 5 oz
Breeds Galápagos Islands, year-round
Winters Seas around breeding area, year-round
Journey length 0–60 miles

ROYAL ALBATROSS
Diomedea epomophora
Wn 120–138 in **Wt** 19 lb 13 oz–21 lb
Breeds NZ, Campbell, Auckland & Chatham islands, Oct every other year
Winters S Pacific between Australia & S America; also in S Atlantic & off S Africa, perhaps circumpolar year-round
Journey length 3,750–9,300 miles

BULLER'S ALBATROSS
Diomedea bulleri
Wn 80–83 in **Wt** 5 lb 5 oz–6 lb 13 oz
Breeds Islands off NZ, Oct/Jan to Jul/Oct
Winters S Pacific as far as W coast of S America, Jul to Oct/Jan
Journey length 600–7,500 miles

BLACK-FOOTED ALBATROSS
Diomedea nigripes
Wn 75–83 in
Wt 6 lb 10 oz–7 lb 15 oz
Breeds Hawaiian Islands & islands off SE Japan, Nov to Jul
Winters N Pacific, Jul to Nov
Journey length 600–6,000 miles

LAYSAN ALBATROSS
Diomedea immutabilis
Wn 76–79 in **Wt** 5 lb 1 oz–6 lb 3 oz
Breeds Hawaiian Islands & islands off SE Japan & W Mexico, Nov to Jul
Winters N Pacific, Jul to Nov
Journey length 600–4,350 miles

SHY ALBATROSS *Diomedea cauta*
Wn 86–100 in
Wt 7 lb 8 oz–9 lb 11 oz
Breeds Islands off NZ, Tasmania & Crozet Island, Sep to Jul
Winters At sea between S Africa, Australia & W S America, Jul to Sep
Journey length 1,250–7,500 miles

YELLOW-NOSED ALBATROSS
Diomedea chlororhynchus
Wn 78–100 in
Wt 5 lb 8 oz–6 lb 7 oz
Breeds Tristan da Cunha, Gough Island & islands in S Ocean, Aug/Sep to Jul
Winters At sea between E S America, S Africa & Australia, Jul to Aug/Sep
Journey length 600–7,500 miles

GRAY-HEADED ALBATROSS
Diomedea chrysostoma
Wn 70–86 in
Wt 6 lb 10 oz–8 lb 4 oz
Breeds Circumpolar on S Ocean islands, Oct to Aug every other year
Winters Circumpolar at sea, Aug to Oct in year following breeding
Journey length 1,250–9,300 miles

SOOTY ALBATROSS
Phoebetria fusca
Wn 78–80 in
Wt 5 lb 5 oz–5 lb 15 oz
Breeds Islands in S Atlantic & Indian oceans, Jul/Aug to May/Jun every other year
Winters At sea between E S America, S Africa & Australia, May/Jun to Jul/Aug in year following breeding
Journey length 1,250–7,500 miles

LIGHT-MANTLED SOOTY ALBATROSS
Phoebetria palpebrata
Wn 72–85 in **Wt** 6 lb 3 oz–6 lb 13 oz
Breeds Circumpolar on islands in S Ocean, Sep/Oct to Jul every other year
Winters Circumpolar at sea, Jul to Sep in year following breeding
Journey length 1,250–9,300 miles

NORTHERN GIANT PETREL
Macronectes halli
Wn 70–78 in **Wt** 8 lb 6 oz–11 lb
Breeds Circumpolar on islands in S Ocean, Aug to Mar/Apr
Winters Circumpolar at sea, Mar/Apr to Aug
Journey length 600–9,300 miles

SOUTHERN FULMAR
Fulmarus glacialoides
Wn 44–47 in
Wt 1 lb 10 oz–1 lb 14 oz
Breeds Antarctica & adjacent islands, Nov to Mar
Winters Circumpolar at sea around N & S Australian & S American coasts, Mar to Nov
Journey length 600–3,400 miles

ANTARCTIC PETREL
Thalassoica antarctica
Wn 39–40 in
Wt 1 lb 2 oz–1 lb 11 oz
Breeds Antarctica, Nov to Jun
Winters S Ocean N to Antarctic Convergence, Jul to Oct
Journey length 600–1,500 miles

CAPE PETREL *Daption capense*
Wn 31–35 in **Wt** 12 oz–1 lb 1 oz
Breeds Antarctic & subantarctic islands & islands off NZ, Nov to Mar
Winters Circumpolar at sea as far N as equator in Pacific, Mar to Nov
Journey length 600–5,000 miles

GREAT-WINGED PETREL
Pterodroma macroptera
Wn 37–39 in **Wt** 1 lb–1 lb 10 oz
Breeds Subantarctic islands, SE Australia & N Island, NZ, Apr to Sep
Winters S Ocean N to subtropics of Atlantic, Indian & Pacific oceans, Oct to Mar
Journey length 600–2,800 miles

WHITE-HEADED PETREL
Pterodroma lessonii
Wn 42–43 in
Wt 1 lb 4 oz–1 lb 13 oz
Breeds Islands off NZ & in S Indian Ocean, Oct to Mar
Winters Circumpolar at sea, Mar to Oct
Journey length 600–6,000 miles

BLACK-CAPPED PETREL
Pterodroma hasitata
Wn 35–39 in **Wt** 14 oz–1 lb 5 oz
Breeds W Indies, Dec to Jun
Winters Tropical & subtropical Caribbean & Atlantic, Jul to Nov
Journey length 600–1,500 miles

ATLANTIC PETREL
Pterodroma incerta
Wn 40–41 in **Wt** 1 lb 2 oz–1 lb 9 oz
Breeds Tristan da Cunha & Gough Islands, Mar to Oct
Winters S Atlantic between S Africa & S America, Oct to Mar
Journey length 600–2,800 miles

MOTTLED PETREL
Pterodroma inexpectata
Wn 29–32 in **Wt** 9–16 oz
Breeds S Island, Stewart & Snares islands, NZ, Oct to Mar
Winters Bering Sea & Gulf of Alaska, Apr to Aug
Journey length 3,000–8,000 miles

SOLANDER'S PETREL
Pterodroma solandri
Wn 37–41 in **Wt** 1 lb–1 lb 3 oz
Breeds Lord Howe & Philip islands off E Australia, Mar to Aug
Winters NW Pacific, Sep to Feb
Journey length 3,000–5,600 miles

KERGUELEN PETREL
Pterodroma brevirostris
Wn 31–32 in **Wt** 12–13 oz
Breeds Tristan da Cunha & islands in S Indian Ocean, Aug to Mar
Winters Circumpolar at sea, Mar to Jul
Journey length 3,000–7,500 miles

KERMADEC PETREL
Pterodroma neglecta
Wn 35–37 in **Wt** 1 lb 2 oz
Breeds Islands in S Pacific, year-round depending on locality
Winters S Pacific & S N Pacific, year-round
Journey length 1,250–3,750 miles

JUAN FERNANDEZ PETREL
Pterodroma externa
Wn 37–38 in **Wt** 1 lb–1 lb 3 oz
Breeds Juan Fernandez Islands, Oct to Jun
Winters N into S & C N Pacific, Jun to Oct
Journey length 2,500–4,700 miles

WHITE-NECKED PETREL
Pterodroma cervicalis
Wn 37–41 in **Wt** 13 oz–1 lb 3 oz
Breeds Kermadec Islands N of NZ, Oct to Mar
Winters NW Pacific N to Japan, Mar to Sep
Journey length 3,000–5,300 miles

COOK'S PETREL *Pterodroma cookii*
Wn 25–25 in **Wt** 6–7 oz
Breeds Islands off NZ, Oct to Mar
Winters N Pacific N to Aleutian Islands, Mar to Oct
Journey length 3,000–6,500 miles

MAS A TIERRA PETREL
Pterodroma defilippiana
Wn 25–26 in **Wt** 6–7 oz
Breeds Juan Fernandez Islands, Jul to Feb
Winters W Pacific N to W C America, Mar to Jun
Journey length 1,250–2,500 miles

BLACK-WINGED PETREL
Pterodroma nigripennis
Wn 24–27 in **Wt** 5–7 oz
Breeds Islands off NZ & E Australia, Nov to Apr
Winters W Pacific N to Hawaiian Islands, Apr to Oct
Journey length 2,800–5,900 miles

STEJNEGER'S PETREL
Pterodroma longirostris
Wn 20–25 in **Wt** 5–6 oz
Breeds Juan Fernandez Islands, Nov to Apr
Winters S N Pacific N to Japan & perhaps California, Jun to Nov
Journey length 4,350–6,000 miles

BLUE PETREL *Halobaena caerulea*
Wn 22–27 in **Wt** 6–8 oz
Breeds Subantarctic islands, Sep to Mar
Winters S Ocean N to subtropical waters off S America, Apr to Aug
Journey length 1,850–5,600 miles

SALVIN'S PRION *Pachyptila salvini*
Wn 22 in **Wt** 6 oz
Breeds Amsterdam, St. Paul, Crozet & Prince Edward islands, S Indian Ocean, Oct to Mar
Winters S Ocean N to S Africa & Australia, Mar to Sep
Journey length 600–4,350 miles

ANTARCTIC PRION
Pachyptila desolata
Wn 22–25 in **Wt** 5–6 oz
Breeds Subantarctic & S Indian Ocean islands & islands off NZ, Nov to Apr
Winters S Ocean N to subtropical waters off S America & S Africa, Apr to Oct
Journey length 600–3,400 miles

SLENDER-BILLED PRION
Pachyptila belcheri
Wn 21–22 in **Wt** 5 oz
Breeds Falkland, Crozet & Kerguelen islands, Oct to Apr
Winters S Ocean, N to 15°S off S America & to S Africa & Australia, Apr to Sep
Journey length 1,500–4,700 miles

BLACK PETREL
Procellaria parkinsoni
Wn 43–47 in
Wt 1 lb 8 oz–1 lb 9 oz
Breeds Islands off Japan, Korea & E China, Mar to Oct; islands off N Island, NZ, Nov to May
Winters N breeders W Pacific S to N & E Australia, Oct to Feb; S breeders W coasts of C & S America, Jun to Oct
Journey length 3,750–6,800 miles

BULWER'S PETREL
Bulweria bulwerii
Wn 26–28 in **Wt** 3–5 oz
Breeds Islands in E N Atlantic & in N Pacific from E China to Hawaii, Apr to Sep
Winters Subtropical & tropical Atlantic, Pacific & Indian oceans, Oct to Mar
Journey length 3,000–5,000 miles

GRAY PETREL *Procellaria cinerea*
Wn 45–51 in **Wt** 2 lb–2 lb 11 oz
Breeds Islands in S Atlantic, S Indian Ocean & off NZ, Feb to Jul
Winters S Ocean N to equator in E Pacific, Aug to Jan
Journey length 1,500–5,600 miles

WHITE-CHINNED PETREL
Procellaria aequinoctialis
Wn 52–57 in
Wt 2 lb 4 oz–3 lb 2 oz
Breeds S Ocean islands, Oct to Apr
Winters S Ocean N to equator in E Pacific, May to Sep
Journey length 1,500–5,600 miles

STREAKED SHEARWATER
Calonectris leucomelas
Wn 47–48 in **Wt** 1 lb–1 lb 3 oz
Breeds Islands of Japan S to E China & Korea, Mar to Sep
Winters W Pacific S to N & E Australia, Oct to Feb
Journey length 3,750–5,300 miles

CORY'S SHEARWATER
Calonectris diomedea
Wn 39–49 in **Wt** 1 lb 4 oz–2 lb 2 oz
Breeds Mediterranean islands, Berlengas, Azores, Canary & Cape Verde islands, Apr to Oct
Winters Atlantic W to Gulf of Mexico, S Atlantic coasts of S America & Africa & S Indian Ocean, Nov to Mar
Journey length 2,500–6,000 miles

PINK-FOOTED SHEARWATER
Puffinus creatopus
Wn 42–43 in
Wt 1 lb 9 oz–1 lb 12 oz
Breeds Juan Fernandez Islands & islands off W S America, Nov to Apr
Winters E N Pacific N to Alaska, Apr to Oct
Journey length 3,750–6,800 miles

FLESH-FOOTED SHEARWATER
Puffinus carneipes
Wn 38–42 in
Wt 1 lb 4 oz–1 lb 11 oz
Breeds Islands in S Indian Ocean & off Australia & NZ, Sep to Mar
Winters Indian Ocean & W N Pacific N to Bering Sea, Mar to Sep
Journey length 3,750–8,000 miles

GREATER SHEARWATER
Puffinus gravis
Wn 39–46 in
Wt 1 lb 9 oz–2 lb 2 oz
Breeds Tristan da Cunha group, Gough Island & Falklands, Oct to Apr
Winters N Atlantic N to Greenland, Iceland & Norway, May to Oct
Journey length 4,350–9,000 miles

BULLER'S SHEARWATER
Puffinus bulleri
Wn 37–38 in **Wt** 12–15 oz
Breeds Islands off N Island, NZ, Oct to Apr
Winters N Pacific N to Bering Sea, Apr to Sep
Journey length 4,350–8,700 miles

SOOTY SHEARWATER
Puffinus griseus
Wn 37–42 in
Wt 1 lb 7 oz–2 lb 2 oz
Breeds S Chile & Falkland Islands, islands off S Australia & NZ, Oct to Apr/May
Winters N Pacific & N Atlantic N to subarctic waters, May to Sep
Journey length 6,000–8,400 miles

HUTTON'S SHEARWATER
Puffinus huttoni
Wn 28–35 in **Wt** 12–13 oz
Breeds S Island, NZ, Sep to Apr
Winters Coasts of E, S & W Australia, May to Aug
Journey length 1,500–4,700 miles

FLUTTERING SHEARWATER
Puffinus gavia
Wn 29–30 in **Wt** 8–15 oz
Breeds Islands off N Island, NZ, Sep to Jan
Winters Seas off SE Australia, Feb to Aug
Journey length 1,850–2,500 miles

BLACK-BELLIED STORM-PETREL
Fregetta tropica
Wn 17–18 in **Wt** 2–2¼ oz
Breeds Circumpolar S Ocean islands, Nov to Apr
Winters S Ocean N to Indian Ocean & to equator in Atlantic & Pacific, May to Oct
Journey length 3,000–5,300 miles

SWINHOE'S STORM-PETREL
Oceanodroma monorhis
Wn 17–18 in **Wt** 1¼–1½ oz
Breeds Japan, China & Korea, Apr to Sep
Winters N Indian Ocean & Arabian Sea, Oct to Mar
Journey length 5,000–8,000 miles

LEACH'S STORM-PETREL
Oceanodroma leucorhoa
Wn 17–18 in **Wt** 1¼–1½ oz
Breeds Islands in N Pacific S to Mexico & in N N Atlantic, May to Sep
Winters Equatorial Pacific Ocean & S Atlantic, Oct to Apr
Journey length 3,000–5,600 miles

BLACK STORM-PETREL
Oceanodroma melania
Wn 18–20 in **Wt** 1½–1¾ oz
Breeds Islands in Gulf of California, May to Sep
Winters E Pacific S to N S America, Oct to Apr
Journey length 600–2,800 miles

MATSUDAIRA'S STORM-PETREL
Oceanodroma matsudairae
Wn 21–22 in **Wt** 2 oz
Breeds Kazan-retto (Volcano Islands) SE of Japan, Jan to Jun
Winters Indian Ocean, Jul to Dec
Journey length 4,350–6,000 miles

SOUTH POLAR SKUA
Catharacta maccormickii
Wn 51–55 in
Wt 2 lb 10 oz–3 lb 8 oz
Breeds Antarctica, Oct to Apr
Winters N Pacific & E Atlantic, May to Sep
Journey length 3,000–9,300 miles

WHITE-CHEEKED TERN
Sterna repressa
Wn 22–25 in **Wt** 4–5 oz
Breeds NZ & neighboring islands, Sep to Jan
Winters S & E Australia, Feb to Aug
Journey length 600–1,500 miles

CAYENNE TERN *Sterna eurygnatha*
Wn 37–39 in **Wt** 7–10 oz
Breeds Coasts of Venezuela S to N Brazil, Mar to Sep
Winters Coasts of S America S to Argentina, Oct to Feb
Journey length 1,250–3,750 miles

Index

Acknowledgments

FURTHER READING

Alerstam, Thomas *Bird Migration* Cambridge University Press, Cambridge, 1990

Baker, R. Robin *Fantastic Journey: The Marvels of Animal Migration* Merehurst, London; Weldon Owen, Australia, 1991

—— *Bird Navigation: The Solution to a Mystery?* Hodder & Stoughton, London, 1984

—— *The Mystery of Migration* Macdonald Futura, London, 1980

—— *The Evolutionary Ecology of Animal Migration* Hodder & Stoughton, London, 1978

Berthold, Peter *Bird Migration: A General Survey* Oxford University Press, Oxford, 1993

—— (ed.) *Orientation in Birds* Birkhäuser Verlag, Basel, 1991

Bub, Hans *Bird Trapping and Bird Ringing: A Handbook for Trapping Methods All Over the World* Cornell University Press, Ithaca, New York State, 1991

Burton, Robert *Bird Migration* Aurum Press, London; Facts on File, New York, 1992

Dorst, Jean *The Migrations of Birds* Heinemann, London, 1962

Elkins, Norman *Weather and Bird Behaviour* T. & A.D. Poyser/Academic Press, London, 2nd ed., 1988

Hagan, James M. III and **David W. Johnston** (eds.) *Ecology and Conservation of Neotropical Migrant Landbirds* Smithsonian Institution, Washington, D.C., 1992

Heintzelman, Donald S. T*he Migrations of Hawks* University of Indiana Press, Bloomington, Indiana, 1986

Keast, A. and **E.S. Morton** (eds.) *Migrant Birds in the Neotropics* Smithsonian Institution, Washington, D.C., 1980

Matthews, G.V.T. *Bird Migration* Cambridge University Press, Cambridge, 2nd ed., 1968

Mead, Chris *Bird Migration* Country Life, London, 1983

Moreau, R.E. *The Palaearctic-African Bird Migration Systems* Academic Press, London, 1972

Pearson, Bruce *An Artist on Migration* HarperCollins, London, 1991

Salathé, T. (ed.) *Conserving Migratory Birds* ICBP Technical Publication No. 12, BirdLife International, Cambridge, 1991

Terborgh, John *Where Have All the Birds Gone? Essays on the Biology and Conservation of Birds That Migrate to the American Tropics* Princeton University Press, Princeton, New Jersey, 1989

PHOTOGRAPHS

2 Alan and Sandy Carey / nature wildlife and the environment; 6 Tony Hamblin/FLPA; 13 Tom Walmsley/Naturepl.com; 15 Frans Lanting/Minden Pictures/FLPA; 19 Jim Brandenburg/Minden Pictures/FLPA; 25 Flip De Nooyer/Foto Natura/FLPA; 27 Desert Wheatear Neil Bowman/FLPA; 29 Konrad Wothe/Minden Pictures/FLPA; 33 Mark Newman/FLPA; 37 Frans Lanting/Minden Pictures/FLPA; 39 Jim Brandenburg/Minden Pictures/FLPA; 43 Frans Lanting/Minden Pictures/FLPA; 44 Snow Goose David Hosking/FLPA; 45 Peter Reynolds/FLPA; 46 Flip De Nooyer/Foto Natura/FLPA; 47 Frans Lanting/Minden Pictures/FLPA; 48 Cyril Ruoso/JH Editorial/Minden Pictures/FLPA; 49 Piet Munsterman/Foto Natura/FLPA; 50 Malcolm Schuyl/FLPA; 54 Tim Fitzharris/Minden Pictures/FLPA; 56 Tom Vezo/Minden Pictures/FLPA; 60 Killdeer Malcolm Schuyl/FLPA; 63 William S. Clark/FLPA; 69 Neil Bowman/FLPA; 70 Tui De Roy/Minden Pictures/FLPA; 77 S & D & K Maslowski/FLPA; 78 William S. Clark/FLPA; 80 Paul Hobson/FLPA; 85 Hannu Hautala/FLPA; 89 Fritz Polking/FLPA; 93 Yossi Eshbol/FLPA; 95 Michael Callan/FLPA; 97 Winfried Wisniewski/FLPA; 98 David Hosking/FLPA; 105 Charlie Brown/FLPA; 110 Frits Van Daalen/Foto Natura/FLPA; 114 Alan and Sandy Carey/nature wildlife and the environment; 119 Thomas Mangelsen/Minden Pictures/FLPA; 120 Roger Wilmshurst/FLPA; 123 Winfried Wisniewski/Foto Natura/FLPA; 126 Fritz Van Daalen/Foto Natura/FLPA; 128 Roger Wilmshurst/FLPA; 130 Malcolm Schuyl/FLPA; 132 Michael & Patricia Fogden/Minden Pictures/FLPA; 135 Martin B Withers/FLPA; 136 Martin B Withers/FLPA; 136 Martin B Withers/FLPA; 136 David Hosking/FLPA; 139 Wendy Dennis/FLPA; 140 Michael Callan/FLPA; 141 David Hosking/FLPA; 142 David Hosking/FLPA; 142 Peter Steyn/Ardea.com; 144 Martin B Withers/FLPA; 144 Neil Bowman/FLPA; 145 Jurgen & Christine Sohns/FLPA; 146 – 00021221.jpg Scarlet Honeyeater NHPA/Daniel Zupanc; 146 Neil Bowman/FLPA; 147 Tui De Roy/Minden Pictures/FLPA; 148 Hiroya Minakuchi/Minden Pictures/FLPA; 153 PhotoDisc, Inc.; 155 Tui De Roy/Minden Pictures/FLPA; 159 Chris Schenk/Foto Natura/FLPA; 161 Eric Wanders/Foto Natura/FLPA;162 Geostock/nature wildlife and the environment.

ILLUSTRATIONS

Norman Arlott
Dianne Breeze
Hilary Burn
Chris Christoforou
Richard Draper
Robert Gillmor
Peter Hayman
Gary Hincks
Aziz Khan
Patrick J. Lynch
Denys Ovenden
David Quinn
Andrew Robinson
Chris Rose
David Thelwell
Owen Williams
Ann Winterbotham
Ken Wood
Michael Woods

MAPS

GoYa! Heidelberg | Madrid
Satellite tracks (pp90) courtesy of Prof. Thomas Alerstam Department of Animal Ecology The Ecology Building Lund University 223 62 Lund, Sweden This map previously appeared in the Journal of Avian Biology-97